ENVIRONMENTAL RESTORATION AND DESIGN FOR RECREATION AND ECOTOURISM

Integrative Studies in Water Management and Land Development

Series Editor
Robert L. France

Published Titles

ENVIRONMENTAL RESTORATION AND DESIGN FOR RECREATION AND ECOTOURISM

Robert L. France

CRC Press
Taylor & Francis Group
Boca Raton   London   New York

CRC Press is an imprint of the
Taylor & Francis Group, an **informa** business

Front cover image is of the Las Vegas Wash moving through Clark County Wetlands Park (Chapters 5, 6 and 9). Small inset images on front cover are of the interpretation center at the London Wetland Centre (Chapters 7, 8 and 9), and of Crissy Fields, San Francisco (Chapter 2). Thumb-nail image on back cover is of a notebook used for planning site development of the London Wetland Centre.

CRC Press
Taylor & Francis Group
6000 Broken Sound Parkway NW, Suite 300
Boca Raton, FL 33487-2742

© 2012 by Taylor & Francis Group, LLC
CRC Press is an imprint of Taylor & Francis Group, an Informa business

No claim to original U.S. Government works

Printed in the United States of America on acid-free paper
Version Date: 2011929

International Standard Book Number: 978-1-4398-6986-4 (Hardback)

Visit the Taylor & Francis Web site at
http://www.taylorandfrancis.com

and the CRC Press Web site at
http://www.crcpress.com

Dedication

For the late Gary Mason,
visionary pioneer of stream daylighting
and impassioned environmentalist.

Contents

Section II: Recovery processes and design practices for regenerating derelict landscapes for recreation and ecotourism

Series statement:
Integrative studies

Ecological issues and environmental problems have become exceedingly complex. Today, it is hubris to suppose that any single discipline can provide all the solutions for protecting and restoring ecological integrity. We have entered an age where professional humility is the only operational means for approaching environmental understanding and prediction. As a result, socially acceptable and sustainable solutions must be both imaginative and integrative in scope; in other words, garnered through combining insights gleaned from various specialized disciplines, expressed and examined together.

The purpose of the CRC Press series Integrative Studies in Water Management and Land Development is to produce a set of books that transcends the disciplines of science and engineering alone. Instead, these efforts will be truly integrative in their incorporation of additional elements from landscape architecture, land-use planning, economics, education, environmental management, history, and art. The emphasis of the series will be on the breadth of study approach coupled with depth of intellectual vigor required for the investigations undertaken.

Robert L. France
Series Editor
Integrative Studies in Water Management and Land Development
Associate Professor of Watershed Management
NSAC
Science Advisor for the Center for Technology and Environment
Harvard University
Principal, W.D.N.R.G. Limnetics

Acknowledgments

I am indebted to the inspirational actions and wise words of the participants involved in the case studies that make up this volume: Robert Gearheart, Jeff Harris, Doug Hulyer, Mary Margaret Jones, Gary Mason, Kevin Peberdy, Mark Raming, Vicki Scharnhorst, Malcolm Whitehead, and Becky Zimmerman. Participants in the Clark County Wetlands Park and the London Wetland Centre are thanked in particular for making available all the illustrative images used in this publication in order to spread the message about these two award-winning projects. Main chapter titles in Part I are derived from Francois Truffaut's film, Annie Dillard's book, and John and Michelle Phillips's song.

Designing new natures for people and ecology: Metaphysics, semantics, logistics

> "There is no need to get back to pristine nature even if we could. On weekends many people sit in plastic and metal boxes to go for two or three hours to have a sandwich or barbecue in 'nature' and then return to their so called 'no nature.' Why do we have a civilization where our dreams are so far away? Why can't we work in a way that our living space has a quality that this is our new nature—our urban nature? "
>
> —**Herbert Dreiseitl, Foreword** *in* **France, R.L.**
> *Deep Immersion: The Experience of Water,* **2003**

Metaphysics: Designing space and time

Nature—culture gradients

Environmental restoration in the broadest sense implies creating new nature. But the word "nature" has often been referred to as being the most complex word in the entire English language. For a half decade I taught a first-year seminar titled *The Invention of Nature,* which addressed the term as both an intellectual concept and as a physical reality. The course began by briefly examining, discussing, and then debating various aspects of the culture of nature: Is nature real, or is it just a cultural construct? How do historical "cultural landscapes" and contemporary "urban wilds" blur the boundaries between humans and nature? Where does "wilderness" lie on the spectrum? And how does a memory of landscape—a "topophilia"—influence our assessments of spaces as places? With this conceptual foundation established, the course then examined the gradient of physically manipulating or designing nature: first, the design *of* nature (artificial natures like computer simulations and constructed simulacra, hyper-natures, and also gardens, parks, and zoos); second, design *with* nature (creation of ecological designs and reclaiming degraded sites); and finally, design *for* nature (ecological restoration and wildlife/wilderness management).

After asking the students to describe the most natural place they had ever been, I would then present a slide show of various images from around the world designed to challenge, confound, and ultimately confuse them as to what might be regarded as "natural": a scree slope tumbling down a mountainside in the English Lake District that is actually composed of the discarded remains from a Neolithic axe factory; a group of beautiful small ponds also in the Lake District that were actually created by human engineering in the nineteenth century juxtaposed with ponds in the boreal forest of Ontario that were similarly artificially formed but this time by beavers; the artificially created West Lake in Hangzou, the most famous lake in China, juxtaposed with a replica of West Lake created for the Emperor at the Summer Palace outside of Beijing; a landscape view of Glendaloch in Ireland where the hillside scree patches were created by early miners; polders in the Netherlands and the moors near Hadrian's Wall in north England and their resulting artificial landscapes created, respectively, through manipulations of hydrology or grazers; mountain and forest landscapes of New England that were manipulated by Amerindians through deliberate burning to aid their hunting; the Sahara and Sahel juxtaposed with goats that are partially responsible for the desertification; the boreal forest of Canada which is now an artificial landscape as a consequence of forest fire suppression in what is one of the largest ecosystem manipulations currently occurring on the planet; and then finally, images from an expedition to one of the remotest places on the planet, Ellesmere Island in the Canadian High Arctic, followed by images of me collecting plants for biomonitoring of the long-range, transpolar transport and deposition of trace contaminates. The point of all this is to demonstrate that Nature (with a capital "N") is as much a cultural construct as nature is a "real" entity (Evernden 1992; Cronon 1996), and that in terms of the latter, all locations can be positioned somewhere along a gradient of culture and nature depending on the duration of human contact (Steedman 2005; Figure 1).

Refuse—Relic gradients

Environmental restoration in the broadest sense often implies repairing damaged, derelict, or wasted landscapes. But the way in which we regard the worthiness and therefore wealth of our waste products, be they objects or landscapes, is itself relative. For the aforementioned

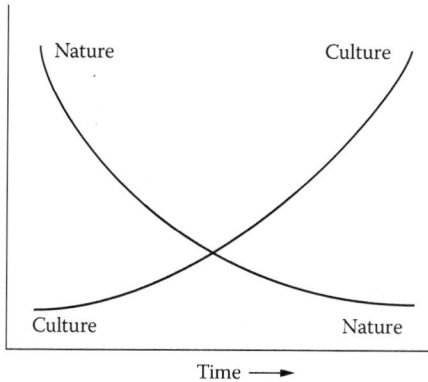

Figure 1 Relative dominance of nature and culture in landscapes along a temporal gradient reflecting human interactions.

course introduction, I would continue challenging the students with more slides, in this case questioning what they considered to be either garbage or valuable resources: an image of a nineteenth-century ghost town in the American southwest that was abandoned but is now protected within a national park and which was used in the famous cycling scene in the movie *Butch Cassidy and the Sundance Kid*; an image of an abandoned, nineteenth-century plague-village in western Ireland that has recently been reinvented as a modern interpretive tourist attraction; close-up photos of the twisted, rusty piping from two derelict industrial complexes followed by distant views of the two locations, which have been transformed into the iconic public recreational spaces of Gasworks Park in Seattle and Emscher Landscape Park in Germany; the High Arctic of Ellesmere Island and paired photos of old barrel rings and broken boxes and building planking contrasted with abandoned gasoline drums and food boxes, the first group being the remains of famous polar expeditions from the past, which are actually referred to as "relics" in a pseudo-religious connotation whereas the latter being "stuff" left behind by modern scientists that is regarded by all as being merely "trash"; a photo of the natural-looking Monte Testaccio in Rome, which is actually a garbage dump composed of the remains of millions of broken sherds of Roman pottery; and finally two similar-looking images from Israel: one of the archeological tell at Biet She'an composed of Chalcolithic- to Byzantine-period remains and the other of the Hiriya landfill, which contains the accumulated garbage of Tel Aviv and is slated to be transformed into a city park. The point of all this is to demonstrate that one Age's garbage can become another Age's artifacts (Rathje and Murphy 2001; Royte 2006) and that the value we impart upon any object or landscape is often a function of time (Light 2008; Figure 2).

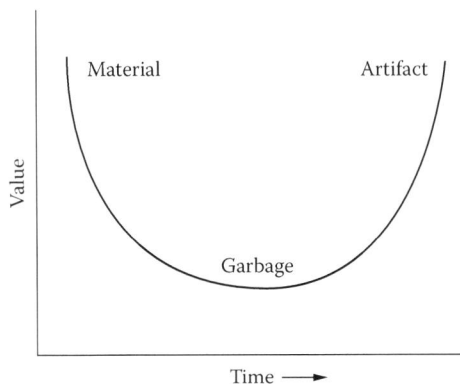

Figure 2 Relative value in objects from being regarded as "material," through "garbage" to being considered "artifacts" along a temporal gradient.

Temporal gradients

Environmental restoration implies a return in time to a predisturbance state. At this point in the course introduction, students usually became so befuddled that further slides were no longer necessary for the last bit of mental knotting. Restoration is the most intellectually intriguing and conceptually challenging of all forms of environmental management. This is particularly the case with issues concerning time and historical fidelity (Hall 2009). We live in a universe where time flows forward linearly. But by creating new natures on old, damaged, or derelict landscapes, restoration attempts to bend time's arrow backward upon itself. It is even more complicated than that. For in order to restore the future, we must be able to predict the past (France 2008a). All becomes tautologically confounded as if in a "back to the future" wormhole (Figure 3). And to add to the confusion, just whose idea of which time should our restorations aspire to? Should we do everything possible in an attempt to replicate a particular instance in time? And why that particular instance and not another, for example? It is really no wonder then that the subject has captured the attention of philosophers (Elliot 1987; Katz 2002; Higgs 2003; Foster 2008; Light 2008; Kidner 2008; Spelman 2008).

Semantics: Designing restoration and regeneration paradigms

The five case studies featured in this book are located in downtown London and San Francisco and on the outskirts of Las Vegas and of a town in northern California. Clearly these urban and suburban locations are remarkably different from those where one might first expect ecological

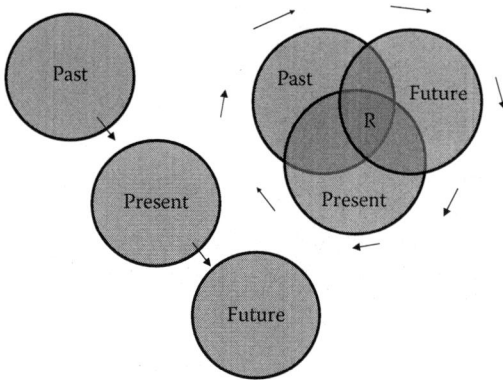

Figure 3 Although time flows linearly forward as shown in the left-hand schematic, restoration ("R" on the right-hand schematic) takes place at the convoluted temporal interface involved with predicting the past in order to transform the present into some re-imagined future.

restoration to take place (Ryan 2008). That is because these case studies are leading examples of a new trend of "environmental reparation with people in mind" that is squarely placed at the interface of nature and culture (France 2008b), and which frequently deals with cities and severely degraded landscapes where a return to pristine conditions may be neither possible nor even desirable (Berger 2008; France 2008c).

Two major distinctions can be made amongst the various subdisciplines of environmental reparation: restoration ecology and restoration design (France 2008b, 2010a). *Restoration ecology* is often preoccupied with historical fidelity (France 2008a, 2008d), the process of ecological restoration being concerned with repairing the damaged bits and pieces of nature, often through improving habitat templates to support restocked or recolonized wildlife. Through *ecological restoration* it may be possible to return a site to its original state. More realistically, it is possible through *ecological rehabilitation* to recover many of the ecological characteristics of a site without a necessary return to its original conditions (France 2010a).

Restoration design, in contrast, is, in the broadest sense, a process by which participants creatively develop physical and conceptual relationships to engage repaired nature through the architectural transformation of their inhabited ecological space, as well as their internal environments (France 2008a, 2008d). In terms of action, transformation of a site to a completely new and artificial altered state for human utilitarian purposes without much attempt to mimic the original natural conditions is the process of *environmental reclamation* (France 2010a). In contrast—and where the case studies of the present book can be situated—the partial recovery of environmental conditions of a site to a new altered state achieved for the mixed benefits of both humans and nature is the process *environmental regeneration* (France 2010a). *Regenerative landscape design* is therefore a multifaceted strategy for improving the living conditions of *both* human and nonhuman nature (France 2008b) and is distinct from other reparative disciplines of environmental engineering, low-impact development, sustainable development, and ecological restoration, perhaps the largest difference being the degree of attention paid to sociological concerns of fostering human–nature interactions (France 2008e; Abbott 2008).

Logistics: Designing a book

Following a strategy that I have used previously (France 2006, 2007, 2010b), this book is based on personal transcriptions made from a series of filmed presentations. It is important to note that in all cases, I have subsequently reworked, updated, and augmented the final chapter texts. Texts for the two comprehensive case studies that form the bulk of the current book—chapters 5 and 7—were sent to the original presenters, who in turn made additions and corrections that I have endeavored to incorporate. Answers to a series of questions—chapters 6 and 8 —were provided by the case study participants either during their original presentations or in response to my later submissions. If mistakes exist in any of the chapters in this book it is a result of my own misinterpretations.

The California case studies covered in this book are three of the pioneering and most inspiring examples of small-scale regenerative landscape design. Today these landscapes exist as extremely popular city parks that attract hundreds of thousands of visitors. The two other case studies detailed in this book—the transformation of the degraded Las Vegas Wash into Clark County Wetlands Park and of the abandoned Barn Elms Reservoirs into the London Wetland Centre—can be regarded as quite possibly *the* most successful, informative, important, and meritoriously accoladed and awarded regenerative landscape design projects that have been undertaken anywhere in the world. Hundreds of thousands of people have visited these two wetland park sites since their creation. These two studies were specifically selected in order to promote the message that the most important regenerative landscape design projects cannot be undertaken without the cooperation and integration of many different disciplines. In other words, regeneration is both far too important and far too complex to remain the purview of any single group of practitioners, such as, for example, ecologists, engineers, or landscape architects.

All case study chapters provide site descriptions, the various actions taken and results ensuing (both structural and nonstructural), the challenges met, and the lessons learned. Additionally, for the two comprehensive case studies, detailed information is provided about the evolution of their regenerative landscape designs. What is most unusual about this book, therefore, compared to other published compilations of case studies, is the emphasis placed on providing the rich backstory to the final design products; in other words, a deliberate focus on the *process* of regenerative landscape design, both through words and, equally as important, through images of designs and plans, some very preliminary in nature. This unusual approach is reinforced by the active voice used in the chapter titles for these two case studies; in other words, use of the words "from" and "to."

Focusing merely on ends (products) while ignoring means (processes) compromises true understanding and makes judgment about projects difficult and possibly superficial. Instead, examining how ideas were first conceived and then later adopted, transformed, or possibly even abandoned along the way, allows readers to follow the complex time course of regenerative landscape design and end-use development, in these cases for recreation and ecotourism. Such detailed attention to history, planning, methodology, and design and construction processes is needed to be able to fully appreciate truly innovative case studies of environmental design such as those to be covered in this book. Yet, oddly, this has rarely been accomplished (for example, Werthmann 2007), at least to the level of detail as covered herein.

Literature cited

Abbott, R. 2008. "Visible cities: A meditation on civic engagement for urban sustainability and landscape regeneration," in *Handbook of regenerative landscape design*, ed. R. L. France, i–v. Boca Raton, FL: CRC Press.

Berger, A. 2008. *Designing the reclaimed landscape.* Boca Raton, FL: Taylor & Francis.

Cronon, W., ed. 1996. *Uncommon ground: Rethinking the human place in nature.* New York: W.W. Norton.

Elliot, R. 1997. *Faking nature.* New York: Routledge.

Evernden, N. 1992. *The social creation of nature.* Baltimore: Johns Hopkins Univ. Press.

Foster, C. 2008. "Wherefore the rhizome?: Eelgrass restoration in the Narragansett Bay," in *Handbook of regenerative landscape design,* ed. R. L. France, 111–42. Boca Raton: CRC Press.

France, R. L. 2006. *Introduction to watershed development: Understanding and managing the impacts of sprawl.* Lanham, MD: Rowman & Littlefield

France, R. L. 2007. *Wetlands of mass destruction: Ancient presage for contemporary ecocide in southern Iraq.* Winnipeg: Green Frigate Books.

France, R. L. 2008a. "Conclusion: Landscapes and mindscapes of restoration design," in *Healing natures, repairing relationships: New perspectives on restoring ecological spaces and consciousness,* ed. R. L. France, 219–26. Winnipeg: Green Frigate Books.

France, R. L. 2008b. "Environmental reparation with people in mind: Regenerative landscape design at the interface of nature and culture," in *Handbook of regenerative landscape design,* ed. R. L. France, ix–xii. Boca Raton, FL: CRC Press.

France, R. L., ed. 2008c. *Handbook of regenerative landscape design.* CRC Press.

France, R. L. 2008d. "Engaging nature and establishing relationships through restoration design," in *Healing natures, repairing relationships: New perspectives on restoring ecological spaces and consciousness,* ed. R. L. France, 7–22. Winnipeg: Green Frigate Books.

France, R. L. 2008e. "Reparative paradigms: Sociological lessons for Venice from regenerative landscape design," in *Handbook of regenerative landscape design,* ed. R. L. France, 427–36. Boca Raton, FL: CRC Press.

France, R. L. 2010a. "Introduction to the paradigm: Ecocultural restorative redevelopment as a guiding principle in rebuilding devastated landscapes in uncertain times, in *Restorative redevelopment of devastated ecocultural landscapes,* ed. R. L. France, 1–7. Boca Raton, FL: CRC Press.

France, R. L. et al. 2010b. *Restorative redevelopment of devastated ecocultural landscapes.* Boca Raton, FL: CRC Press.

Hall, M. 2009. *Restoration and history: The search for a usable environmental past.* New York: Routledge.

Higgs, E. 2003. *Nature by design: People, natural process, and ecological restoration.* Cambridge, MA: MIT Press.

Katz, E. 2002. Understanding moral limits in the duality of artifacts and nature: A reply to critics. *Ethics and the Environ.* 7:138–46.

Kidner, D. 2008. "Nature's memory: Restoration and the triumph of the cognitive," in *Healing natures, repairing relationships: New perspectives on restoring ecological spaces and consciousness,* ed. R. L. France, 69–94. Winnipeg: Green Frigate Books.

Light, A. 2008. "Restorative relationships from artifacts to "natural" systems," in *Healing natures, repairing relationships: New perspectives on restoring ecological spaces and consciousness,* ed. R. L. France, 95–116. Winnipeg: Green Frigate Books.

Rathje, W., and C. Murphy. 2001. *Rubbish! The archeology of garbage.* Tuscon: Univ. Arizona Press.

Royte, E. 2006. *Garbage land: On the secret trail of trash.* New York: Little, Brown & Co.

Ryan, R. 2008. "Understanding the role of environmental designers in environmental restoration and remediation," In *Healing natures, repairing relationships: New perspectives on restoring ecological spaces and consciousness,* ed. R. L. France, 199–218. Winnipeg: Green Frigate Books.

Spelman, E. 2008. "Embracing and resisting the restorative impulse," in *Healing natures, repairing relationships: New perspectives on restoring ecological spaces and consciousness,* ed. R. L. France, 122–139. Winnipeg: Green Frigate Books.

Steedman, R. J. 2005. "Buzzwords and benchmarks: Ecosystem health as a management tool," in *Facilitating watershed management: Fostering awareness and stewardship,* ed. R. L. France, 17–24. Lanham, MD: Rowman & Littlefield.

Werthmann, C. 2007. *Green roof gardens: A case study.* New York: Papress.

About the author

Robert L. France is associate professor of watershed management at Nova Scotia Agricultural College (NSAC). Dr. France has conducted research in regions from the High Arctic to the tropics, on subject areas from bacteria and algae to whales, as well as on chemistry and environmental theory. He has taught at the universities of McGill, Ca'Foscari Venice, and Harvard. France is an acquisition editor for CRC Press, where he runs the Integrative Studies in Water Management and Land Development series, and is also on the editorial board of the independent environmental press Green Frigate Books. He has published over two hundred articles and is the author or editor of over a dozen books of both a technical nature as well as general public interest. France conducts research on the environmental restoration of postagricultural and postindustrial landscapes, integrated watershed management and water-sensitive planning and design, the use of stable isotope analysis to trace material flow in aquatic foodwebs, the impacts of clear-cutting on land–lake linkages, landscape modifications at the suburban-agricultural interface, agricultural urbanism, environmental biography, and immersion into historic agricultural and utilitarian landscapes.

Sources

Clark County Wetlands Park

Becky Zimmerman is a landscape architect and president of Design Workshop. She specializes in tourism, community, economy and market development, and brownfield redevelopment.

Vicki Scharnhorst is an engineer and manager with Kennedy/Jenks Consultants and was formerly Senior Vice President at MWH. She specializes in aquatic and environmental restoration, sustainable development, and project management.

Mark Raming is a wildlife biologist and Executive Vice President at SWCA Environmental Consultants. He specializes in environmental inventory and planning analysis and natural resource management.

Jeff Harris is a planner and the Director of Flathead County Planning and Zoning and was formerly the Manager of Planning and Fiscal Services at Clark County Parks and Recreation. He specializes in open-space planning and environmental project management.

London Wetland Centre

Malcolm Whitehead is a conservation biologist and head of Discovery and Learning at the Zoological Society of London and was formerly the Education and Visitor Services Manager at the Wildflowl and Wetlands Trust (WWT) London Wetland Centre. He specializes in biodiversity conservation, environmental education, wildlife tourism, and exhibit design.

Doug Hulyer is a conservation biologist who sits on the Board of Natural England and is Chair of the Education and Public Understanding Group for the England Biodiversity Strategy, and Chair of the Wetland Vision for England Partnership, and was formerly Director of Conservation Programs at the WWT and Project Director of the London Wetland Centre. He specializes in wildlife conservation, environmental education, and sustainable development.

Kevin Peberdy is a wetland scientist and Director of Centre Development for Wildflowl and Wetlands Trust (WWT). He specializes in wetland design, hydrology management, waterscape ecology, and waterfowl conservation.

California Case Studies

Robert Gearheart is a biologist and environmental engineer and is an Emeritus Professor of Environmental Engineering at Humboldt State University. He specializes in wastewater treatment and water management in the developing world.

Mary Margaret Jones is a landscape architect and President/Senior Principal with Hargreaves Associates. She specializes in urban park design, public space planning and design, and project management.

Gary Mason was a landscape architect and President of Wolfe Mason Landscape Architects and Environmental Scientists. He specialized in public park creation and environmental restoration.

Brown fields and gray waters: Creating public greenspace from regenerated marginal landscapes

chapter 1

Day for night
Stream daylighting in the
San Francisco Bay area*

Water is the base element that sets landscapes in motion, making them feel alive and infusing them with poetry. Water used to be treated as an essential element, as an important part of our communities. Water is precious, but what have we done to it? We started to strangle it in the past century; we turned it into channels; we tried to make it an evil; we treated it with neglect. In short, we have treated our landscapes as waste receptacles, dumping water out of pipes and creating eroded hillsides. And we insist upon burying water underneath roadways wide enough to land airplanes on. We need to stop making the car the prime factor driving community planning. No more should the first questions in planning meetings be about how much parking or how many streets there will be.

Recall your childhood favorite place of water, where you lived and where water first came into your awareness as a source of pleasure. This is what has been lost in most cities. We have made water a feared element, such that people are now afraid of it. Stream restoration projects face this challenge all the time with the public thinking that the water is all of a sudden going to take their children and sweep them away to the ocean. This fear is particularly the case when once-buried streams are opened up through the process of *daylighting*. In such cases, people think that rats are suddenly going to come out of a culvert and take over their neighborhood or that the homeless are all going to live in these restored areas.

Such attitudes are a shame because water is a valuable element whose form can be used to create important civic places and whose function is critical for purposes of green infrastructure. It is therefore critically important to begin to think of water as a resource rather than a problem. This amounts to a significant mindset change for many city engineers and planners in terms of water management. These individuals have typically regarded water as something to "get outta town" as quickly as possible. In reality, the biggest problem of flooding is not the water but ordinances that enable people to live where they shouldn't and allow them to build too close to the river or stream.

---

* Adapted from a presentation by Gary Mason.

The good news is that many ecologists, hydrologists, and landscape architects are fighting back. A whole cadre of people are now bringing back nature into cities. And one high-visibility approach that is gaining great attention in this movement is stream daylighting. A report by Rocky Mountain Institute that compiled and chronicled such projects around the United States determined that the San Francisco Bay area was one of the birthing places of this process. Daylighting originated there in the early 1980s with people involved in park and open space design. This chapter will review four past and a potential future project through the lenses of practicalities, community wishes, and design approaches.

Daylighting is a component of stream restoration wherein buried streams are returned to the landscape. It is very important to recognize that it is not a process of bringing the water up to the surface of the land but rather a process of reallowing the stream to create its normal stream profile in the landscape, which means it is necessary to go down to the water. This process requires designers to think three-dimensionally.

The first task is to find the water, and here the first take-home message is that water cannot be controlled but must be worked with. The second is if a stream is to be restored, it has to meet certain criteria; restorers cannot simply dig it up with a backhoe or an excavator and let the water run through and expect a restored stream. Rather, it is necessary to design a restored stream back in our communities in a way that's safe for people and works within the urban landscape. It is essential that a detailed scientific and engineering understanding be in place before the project begins. One rule of fluvial geomorphology is that a restored stream has to have a floodplain. A restored stream also has to have a dynamic equilibrium between erosion and sedimentation, a balance between the amount of sediment picked up and deposited along the channel. A restored stream has to support aquatic and terrestrial wildlife, and it has to have a curvature that is appropriate to its width, slope, and bed material. If these components do not exist, it is not a stream restoration, though it might be a water feature. In the end, to truly restore a stream it is important to get back into the ecosystem of the surrounding landscape and not only recognize but understand how the landscape influences stream functions. It is important to remember that a restored stream is not an engineering problem; in other words, we are not interested in catching the water and quickly moving it away. Just the opposite, in fact: You want to catch it and slow it down. Stream restoration and daylighting projects are terrific ways to do this in the community.

Stream daylighting provides several practical benefits including reducing flooding and relieving choke points. In many circumstances, stream restoration and daylighting projects are held to a higher standard than pipes, such as a 100-year versus 25-year flood capacity. As such, daylighting actually helps stormwater management by taking water out of the smaller pipe and opening up a big channel that has a higher capacity. Also, daylighting is almost always a less expensive alternative to separating combined sewers, as well as to replacing deteriorating culverts and other crumbling pieces of gray infrastructure. Further, it is very important that the improvements in water and habitat quality resulting from daylighting are brought about through adherence to the Clean Water Act.

Daylighting provides value-added community benefits as well. When water is placed into a pipe, the result is water in a pipe. The water is out of sight and out of mind, and there are no functional benefits. But in daylighting, the values include recreation trails and greenways, outdoor classrooms, community building through connections to nature, as well as a neighborhood amenity. Daylighting therefore serves to connect people physically to the natural systems in which they live. Daylighting projects are like scavenger hunts in that those working on the project must find out where the water was, where it is now going, how it got shunted off into different pipes and culverts, and how the system can be made to work again. It is therefore a fascinating game that easily gets people excited. Such projects translate science, engineering, and design into a common language that people understand. Daylighting is also important in that it allows people to see how waterways can become integrated into the vital structure of the city, not just as a nice extra but rather as an essential element worthy of obtaining an essential part of the city budget.

People do get excited about daylighting. When the project manager of the first such project in San Francisco visited the site that was to be converted into the Suds Flatlands Park ("Suds" being an acronym of the four bordering streets) he saw the culvert at one end where the stream—Strawberry Creek—was open upstream. And so on the flyer to announce the first public planning meeting he referred to the site as "Strawberry Creek Park." People came to the meeting having no idea about any such creek nor where a park of such a name would actually be located. The brilliant response to their inquiries about the location to the mysterious Strawberry Creek Park was: "It is 15 feet down and waiting...," which completely captured the imagination of the people. The public then rushed out to the site eager to be shown how the stream had been culverted and how it could be resurrected, eventually leading to the first daylighting project in San Francisco and one of the first in North America. The lesson is never to underestimate the power of a simple inspirational challenge in the process of allowing water to be reborn.

Begun in 1984 with the city of Berkeley as the client, the daylighting of Strawberry Creek from its internment in an abandoned 4-acre rail freight yard has won national awards. Although the headwaters of Bay Area streams are grassland hills, Berkeley and Oakland are densely urbanized areas with little room for the streams to wiggle their way down to the ocean. In consequence, all these projects faced the challenge of urban hydrology. When a comment was made that the park was "15 feet down and waiting," it captured the interest of the planning and parks and recreation commissions but not the public works department who wanted to kill the project right from the start. These city engineers thought the project was unsafe and crazy, inquiring incredulously about why anyone would want to take the stream out of a pipe. Luckily the design team stacked the first meetings with some big-name environmentalists, who were able to lobby the cause so convincingly as to reach near-unanimous approval.

Today the floodplain functions as a public park; during special events, it can double as an amphitheater in which spectators sit on grass slopes. Because of limited funds for stream restoration ($85,000 out of the $400,000 total park budget), the designers were lucky that the culvert they were breaking apart in order to release the stream had not been constructed with rebar (because of its age). Further savings resulted through reusing blocks from a broken-up roadway for stream stabilization.

Within one year, what had begun as a derelict field of concrete had been transformed into a lush, green oasis. Eventually, a canopy of trees developed to shade the picnic tables. And important for dispelling a big fear about daylighting, children have been playing in the forested riparian park for almost two decades without a single instance of safety problems ever having been reported about anyone getting hurt and having to go to a hospital.

Another project, Codornices Creek, is a 400-foot stretch of a culverted creek that was daylighted in 1994 for the Urban Creeks Council in the cities of Albany and Berkeley. This is a project characteristic of those in dense urban areas in which small concentrated sections of daylighting become possible when a vacant area provides an opportunity to grab hold of a buried stream and bring it up to the surface. In such space-constrained situations, it becomes especially important for projects to have the correct hydrology, geomorphology, and engineering. In terms of the latter, it becomes critically important to bring engineers on side in terms of agreement by giving them a different set of working criteria than what they are normally used to. Once informed that for this particular project water must move through the area safely without a pipe, engineers generally will accept the challenge and work toward trying to solve any problems. If their normal water management solutions are taken away, engineers soon enthusiastically participate in daylighting.

The project area was a vacant lot located behind a warehouse that was completely covered over and into which some small feeder drainage pipes drained. Although the warehouse company had wanted to convert the site to an expanded parking lot, a recently passed creek ordinance in the community based on the presence of a 5-foot section of exposed open water on the property meant that permission was denied for burying this remaining section of the stream in a culvert. The ordinance also stated that there needed to be a 30-foot buffer zone on either side of the creek. In the end, the company was forced to relinquish the land for restoration rather than further development. About 400 feet were exposed and restored for a meager $20,000 through thousands of hours of volunteer pick and shovel work to build out the floodplain, as well as a construction worker who moonlighted from his job with a backhoe. Various bioengineering techniques were tried, but many were washed out by a large storm, so the site became a good testing area. The old culvert was kept in place as the foot for the rebuilt streambank with the water now running in the parallel open channel. Fascines and brush were used so that today the banksides are lushly vegetated and the stream channel looks completely natural.

The accompanying streamside park that was created is now used by a neighboring daycare center as a streamside learning area. It is truly remarkable that such a marginal, low-budget place has become such an incredible landscape for the people. Another lesson from this project is that nature is not necessarily beautiful to everyone. People generally like their nature to be clean and green; the concept of its being wild and natural is something that has become foreign to urban dwellers. It is often necessary, therefore, in stream daylighting and restoration projects to educate against such aesthetic prejudices.

Built in 1995 in an existing 60-acre school park for the Berkeley United School Division and the Urban Creeks Council, Blackberry Creek is another award-winning daylighting project. Opportunity was created by earthquake damage and seismic retrofitting to address the creek, which had been culverted in about 1950 behind the old school. A small, concentrated 400-foot section of the stream was brought out of the culvert to alleviate neighborhood flooding and to create a new grass play area bordering the riparian zone. In urban areas where uses are very concentrated, edges must be clearly defined. In the case of this project, the pathway became the border between the clean green of the lawn and the more natural wooded riparian planting; this also allowed disabled access down to the stream where a small learning amphitheatre area was created.

Many concept drawings were required to convince people that the project was nothing to be fearful of. It was also necessary to physically take people out for a site visit. Many people have a hard time visualizing slope—how steep the new site is going to be or how deep the excavation will be. Once in the field, a tape measure demonstrated that a 2:1 or 3:1 slope would not result in the Grand Canyon that some feared. As for many urban greenspace projects, the biggest challenge was that many people wanted the site to look like the mature forested garden shown in some earlier design sketches and not the more natural look that is ecologically more significant.

The site had to be completely dewatered before work could commence. A problem developed when the design team found out that the restored stream needed a wider meander zone and wider width than what was available because of a sacred redwood tree around which tai chi practitioners meditated. In consequence, some reinforcement had to be used on several bends and the riffle run sequence had to be built back in order to give the stream its natural profile in the constrained location. Vegetative growth was rapid, such that by opening day the willows were already starting to grow and children were already playing at the stream edge.

Another project, Baxter Creek, was a community-driven solution to an engineering problem. Built in 1997, the site was in the middle of a neighborhood, and the original idea had been to simply open up the pipe, which was underneath a narrow greenspace sandwiched between two streets. The initial engineering firm had put in a standard engineering solution with large, expensive blocks and a longitudinal straight run. After the first rainstorm, the project failed so that the firm was called back to realign the stream in relation to appropriate hydrology and geomorphology. Daylighting projects are much more than simply digging a trench and throwing the water into it. They must be built in the correct shape, working with the water and using bioengineering to vegetate the banks. A requirement for enhancing aquatic and terrestrial habitat is the development of a full canopy, which is not achieved until after the first half-decade of growth during which time the willows appear messy and scrubby. It is often necessary to tell people that this "adolescent period" is only a temporary stage. After a few years, the riparian corridor and stream look completely natural, and it is almost impossible to imagine that the place was once a blank green lawn.

Ever since the acclaimed success of the Strawberry Creek Park project in the early 1980s, people have wanted to open up more of the creek in downtown Berkeley. In 1998, the city hired a team to conduct a feasibility study on opening up sections by developing a series of simulations of varying degrees of restoration on a block-to-block basis. One proposal was for a full-flow restoration that would meet all the criteria mentioned previously with a natural meandering stream raised up from under one side of a wide street. Another proposal was for a partial-flow restoration that would not meet all the criteria and would be without the full amplitude of meanders and rather a controlled channel. The canal-restoration option would be more like a water feature with most of the flow remaining in a pipe but with some symbolic water running again on top. The final option of painting a blue stripe on the sidewalk was deemed not acceptable to the community.

When visual preference surveys are undertaken in different communities around the United States, the number-one image that people choose for what sort of place they want in their life is taken from a "rails to trails" calendar from Michigan showing a mother cycling beside her child along a forested path. What people want, therefore, is a safe place in nature for their children for recreation. Stream daylighting is an ideal way in which to stop thinking of cities as being in one place and nature as being someplace else. Nature, stream daylighting shows us, can be simply everywhere.

Further reading

Bucholz, T., and T. Younos. 2007. Urban stream daylighting: Case study evaluations. *Virg. Water Resour. Res. Cen.* 23:65–78.

Clark, S., and A. Middleton. 2007. "A multifaceted, community-driven effort to revitalize an urban watershed: The lower Phalen Creek project," in *Handbook of regenerative landscape design*, ed. R. L. France, 61–72. Boca Raton, FL: CRC Press.

Conradin, F., and R. Buchli. 2007. "The Zurich stream daylighting program," in *Handbook of regenerative landscape design*, ed. R. L. France, 47–60. Boca Raton, FL: CRC Press.

Dreiseitl, H. 2003. "Foreword two: Waterworks," in *Deep immersion: The experience of water*, ed. R. L. France. Winnipeg, Manitoba, Canada: Green Frigate Books.

France, R. L., ed. 2005. *Facilitating watershed management: Fostering awareness and stewardship.* Lanham, MD: Rowman & Littlefield.

France, R. L. 2006. *Introduction to watershed development: Understanding and managing the impacts of sprawl.* Lanham, MD: Rowman & Littlefield.

France, R. L. 2007. "Reparative paradigms: Sociological lessons for Venice from regenerative landscape design," in *Handbook of regenerative landscape design*, ed. R. L. France, 427–35. Boca Raton, FL: CRC Press.

France, R. L. 2007. "Swamped! A tale of two restorations: Part I: The view from home," in *Healing natures, repairing relationships: New perspectives on restoring ecological spaces and consciousness*, ed. R. L. France, 1–6. Winnipeg, Manitoba, Canada: Green Frigate Books.

O'Neill, K., and P. Gaynor. 2007. "Retrieving buried creeks in Seattle: Political and institutional barriers to urban daylighting projects," in *Handbook of regenerative landscape design*, ed. R. L. France, 73–110. Boca Raton, FL: CRC Press.

Pinkham, R. 2000. *Daylighting: New life for buried streams.* Boulder, CO: Rocky Mountain Institute

Pinkham, R., and T. Collins. 2002. "Post-industrial watersheds: Retrofits and restorative redevelopment (Pittsburgh, Pennsylvania)," in *Handbook of water sensitive planning and design*, ed. R. L. France, 67–95. BocaRaton, FL: Lewis Publishers

Resh, V. H. 1995. Stream daylighting in Berkeley, CA creek. *Watershed Protect. Tech.* 1:184–87.

Richman, T. "Innovative diffusion and water-sensitive design," in *Facilitating watershed management: Fostering awareness and stewardship*, ed. R. L. France, 45–51. Lanham, MD: Rowman & Littlefield.

Riley, A. L. 1997. *Restoring streams in cities: A guide for planners, policymakers, and citizens.* Washington, DC: Island Press.

chapter 2

For the time being

Designing a palimpsest landscape
at Crissy Fields, San Francisco*

San Francisco's Presidio is a decommissioned army base that has been converted into a national park situated along some of the most valued coastline within the city. There has been much controversy and debate over the years about how much of the site should become the national park and how much needs to be economically self-sustaining. The Crissy Field site consists of 100 acres and about a mile of bay edge at the foot of the Golden Gate Bridge and is separated from the Presidio by a major highway. There has never been any controversy about its conversion into a national park, which partly results from the site's status as an historic landmark dating back to the 1920s when biplanes landed on the grassy airfield that is itself named Crissy Field. In 1995, Hargreaves and Associates were hired to begin planning how this piece of land might become a national park. Public process in the Bay Area is all-important, and in this case, it lasted for more than 2 years and involved over twenty reviewing agencies. The development of Crissy Field as a national park was a very complex project that included multidimensional, multiagency, multiclient reviews and lots of passion within the public forum.

The site that is now Crissy Field was originally part of the water system of the Bay; it consisted of a wetland that was filled for the 1911 Exposition and was used as a racetrack until the 1920s, at which time it was converted into a grass airfield. In 1995 when the project began, the site was 85 percent asphalt inside, around which were miles and miles of chain-link fence. A few remnant palm trees were scattered about the parking lots from the days when the site had been an army base. Although the site was characterized by abandoned barracks and leftover land, it still offered a great bayside setting and contained a few fantastic areas, such as the remnant dune fields. There were also various restricted zones ranging in type from a thin strip at the water's edge that was protected as habitat for the western snowy plover to other areas where lead paint had been dumped by the army. Before the site could be regenerated as a public park, the army had to go through a cleanup process involving digging up

* Adapted from a presentation by Mary Margaret Jones.

toxic areas and laying the material out in the sun to cook and be heavily irrigated in order to leach out the contaminants before the soil was put back in place.

The first task the team undertook was to study the historic nature of the site. They soon became aware that some vocal members of the public were very interested in turning the entire site back to its original tidal wetland status. The first thing the team realized when consulting old maps was that not only Crissy Field but also the marina district as well as the area beneath the Palace of Fine Arts had all been tidal wetland at one time. The designers quickly realized, therefore, that it would be impossible to go backwards in terms of historical fidelity. The project could not be an ecological restoration but rather would have to proceed toward a new direction.

The designers then investigated the historical timeline of the site, such as being the first racetrack for the exposition, which had been turned into the grassy airfield by the army in the 1920s, which then was later converted into the paved airstrip that was still present. Close examination of the site revealed traces of all these past presences on the landscape. Ultimately the team of designers became interested in letting all the layers of history continue to be read on the site, thereby allowing multiple interpretations of its embodied past. Such an approach was new for the National Park Service, which has one division responsible for cultural resources and another for natural resources. Unfortunately, review of historic maps led to the two camps crystallizing their own respective interests: natural resource staff cared only about restoring the tidal marsh, whereas cultural resource staff cared only about restoring the airfield. Indeed, the latter specifically cared about restoring the airfield to one single year of significance, which in this case that was 1927, what the design teamed coined as "the magic period." But given that all the iterations of the airfield had occurred within a narrow window of only a decade, the design team found that such a mindset focusing on a single year was both bizarre and troublesome.

The design team then put these visions together on a single plan to indicate that neither group could have everything it wanted because of the overlap. The team next created an alternative vision that they called the "yours and my plan," which suggested that one camp could have the airfield at one location and the other camp could have the marsh at another location. In this alternative, both elements could exist side by side, but in the end neither camp would really get what it wanted, and the public probably wouldn't get what it wanted at all. At this point, the public process was initiated.

The first public meeting was attended by three hundred people, and the design team was excited by the turnout, until it soon became obvious that much of the public were angry and were already yelling at the team. Though a diverse group, the most vocal component of the public turned out to be the dog-walkers. These individuals had heard that Crissy Field was going to be converted to a tidal marsh based on recommendations in an earlier National Park Service feasibility study. In the midst of this verbal barrage, design team members began looking in the audience for friendly faces and soon recognized an individual who was a local landscape architect. They called on her and were happy until she stated that, in fact, she walked her dog there every day and really thought the site was just fine as it was. It turned out that there was a large group of people who liked the site specifically because of its leftover status because they could use it however they pleased. Indeed, several individuals even said that they actually liked the chain-link fences because it prevented their dogs from running out into the busy road. Another vocal group, the sailboarders, championed the site as being the fifth-best place in the world for windsurfing, such that it was host to several important competitions. Their goals were to be able to launch from the beach and also to be able to park right next to where they launched. This was because—and they brought along a 70-year-old member to prove their point—some of their club were aged and could not walk a great distance from car to beach. This was humorous given the reality, of course, that it takes a great deal of strength to be able to windsurf in the strong winds that were purported to be present at the site.

In an attempt to accommodate as many different uses as possible, the team came up with a palimpsest strategy wherein the past registered the site's overlayering of history and enabled the mixing and combining of different programs. A series of diagrams were developed to show varying levels of an aggressive approach to mixing the airfield and the tidal marsh, such that the airstrip would be drawn into the marsh and so would become an overlook with fragments left within the tidal marsh itself and, reciprocally, the marsh might reappear again in the airfield. The design team also took some very aggressive approaches to changing road patterns, such that the tidal marsh could be enlarged, including one alternative which would let the road stay in place but convert it into a causeway over an expanded tidal marsh. The problem was that most of the plans with the very large tidal marsh removed the commissary and other buildings important for generating the revenue that the Presidio needed in order to satisfy Congress's demand that the site be economically self-sustaining. In consequence, these buildings had to stay, and the size of the marsh would need to be scaled back accordingly.

The next element that had to be addressed in the planning process was the type of tidal marsh that would be created. To accomplish this, a large team was assembled comprised of an hydrologic engineer; a wetland restoration expert; civil, structural, and marine engineers; and biologists. The key challenge the team faced was how to design the correct *tidal prism*. A tidal prism concerns the shape of the volume of water in the tidal marsh, which enables the tides to move in and out successfully and for the marsh to flush itself based on the mouth staying open and free of sediment clogging. Yet people wanted a very romantic tidal marsh with lovely colored vegetation swaying in the wind, which unfortunately is the sort of system that does not make for a very big tidal prism. Bigger tidal prisms, in fact, occur in marshes, and these are composed mostly of mudflats. As a result, the challenge the design team then faced was the long struggle to convince the public of the beauty of mudflats and to educate them that a tidal marsh does not equate with mosquitoes because of the flushing nature of such systems. Much time was spent engaged in these types of discussions and in trying to achieve a design balance between the bottom part of the tidal prism, the mudflats, and the upland parts of the marsh—the vegetation.

The integrated plan was ultimately approved and progressed through contract documents. Agreed-upon elements included the implied airstrip that would become the overlook into the marsh situated near the main part of the Presidio, the parade grounds of the Presidio extending into the tidal marsh to become an amphitheater and a place for people to look out over the marsh, and parking for windsurfers on reinforced grass cells.

All the asphalt on the site was reused by being ground up and placed underneath the pathways, and all the dirt dug to create the tidal marsh was used to create a series of landforms in the shape of dune-like, wind-swept forms across the site that would help to divide areas and to buffer views and noise from the adjacent freeway. The core design idea for the airfield was to express it as a three-dimensional object in the landscape so that it would be a flat table flush on one end with surrounding grade, and at the east end at the marsh have it become a stage for the cultural life of San Francisco, at the same time as being something that could be pointed to as an object in the landscape for interpretive purposes.

These concepts remained an issue of contention among the cultural resources staff at the National Park Service, who persisted instead that the airfield needed to be restored to its exact condition in 1927, including killing the grass in the exact pattern as the taxiing airplanes had 70 years before. The design team argued that the idea of wanting a truck to drive across the site every day to kill the grass in such a fashion would be antithetical to creating any type of public space. Ultimately, the cultural resources staff were forced into a condition of accepting the layering of all these periods of history because it was really the only way to satisfy all the programmatic desires of the public.

Thus the airfield became the place where people could run their dogs without leashes, and the wetland became a place where dogs were excluded. The snowy plover habitat area was also off-limits to dogs, and the windsurfers were able to park right next to the beach but had to do so on reinforced grass cells. In the end, all this combined to create a place that a wide variety of people could use and hopefully come to eventually love.

Although all the landfill issues were thought to have been sorted out before construction, as soon as excavation work began in the marsh, a series of artifacts were found which halted construction for months while they were assessed by archeologists. As described in the introduction to this book, what they found was "stuff" (bottles, harnesses, belts, rakes, etc.) left behind by the army, which had once been "material" which was then buried because it had been "garbage," but now was considered to be "artifacts" that had to be collected, catalogued, and taken offsite for careful storage.

Construction was halted next when a previously unknown *midden*— which is a different kind of trash from prehistoric Amerindians—was discovered. The leaders were called in from the tribes who might have existed on this site in history, and they then had to decide how to treat the midden. Was the midden simply trash, or was it related to burials? Regardless, it had significant meaning to the Amerindians and so it took some time for them to decide how they wanted it to be treated. Ultimately, it was determined that the material couldn't be excavated, that it had to be covered, and that no one could walk on it. As a result, this area became something different from what had originally been planned. A scrubby dune field resulted that is a very beautiful transition from the grassy dunes to the actual beach dunes through which a boardwalk runs.

The location of the marsh tidal mouth became a hotly contested issue. The first design had situated it right in the middle of the sailboarders' launching area and had to be sited elsewhere so as not to take away their beach. The smaller-than-originally-planned surface area of the marsh posed a challenge for constructing the tidal prism necessary for self-flushing. All the stakeholders had to come back to the table again to decide where the marsh had to grow in size, a process that required negotiating with the cultural resources staff. Eventually these discussions resulted in a change in the nature of the marsh, making it have steeper sloped edges with less gradation of grasses and upland barrier vegetation, and a larger area of mudflats.

Following 6 years of consultation and construction, Crissy Field Park opened in 2001. Many people are still pushing for the removal of the commissary and other buildings in order to expand the tidal marsh, something which is part of the future plan and may eventually be undertaken.

Restoring the dune system was really about putting some structures in place that will evolve over time. Hundreds of volunteers were engaged in helping to plant the dune fields. Remarkably, so many people donated money, from big foundations providing millions to fourth-grade classes giving a dollar, that the surplus capital could be used for maintenance.

Today the park is an amazing place, especially in the late-evening sun, with views of the majestic Golden Gate Bridge behind and of the restored beach and dunes leading into green strip, which leads into the marsh which in turn frames the downtown city center. The architectural shapes of grass ridges and grid layout of trees cast wonderful shadows over the grass field, which is bisected by walkways.

Visitors enter the site along a walkway that provides the urban connection. The first thing that is seen is a grove of cypress trees. Because every gate to the Presidio is marked by such a grove, the team thought it important to remember that they were really constructing a new gateway. This planting design was also hotly debated because such trees might not have originally existed on this site. The design team argued, however, that it was important to recognize such a layer of cultural history from when the site had been the army base.

The landforms screen the view so that visitors do not see the water all at once or right away. Rows of trees draw eyes to the Golden Gate Bridge, and moving past the grass field, visitors arrive at the promenade, which is focused out to the water and back to the city. The landforms also provide places of refuge and prospect. In terms of the latter, the sailboarders had wanted a high landform that could be used as a lifeguard post on which to stand and keep an eye on each other. These individuals are able to set up their expensive equipment on grass beside the bike and walking path.

The promenade crosses over the marsh mouth via a bridge. The mudflats look lovely as the water recedes in complex patterns, and the dunes look completely natural sporting their little tufts of grass. A couple of groves of existing cypress trees were retained at the edge, even though many people wanted them removed because they would not have existed there in a tidal marsh system, but as mentioned, they were culturally an interesting overlay the designers felt important to preserve.

The green lawn of the airfield has become a forum for play and for dog running. On opening day, thousands flocked to the site and filled the sky with hundreds of brightly colored kites. The grass used was native so that its habitat value is high, and its mowing and water requirements are low. At the very end of the site, almost directly underneath the Golden Gate Bridge, a series of intimately scaled, louvered spaces were created that double as a sort of amphitheater with stone steps interfacing with grass landforms for spontaneous performance and observation when sitting on top.

The new seawall defines the old sea edge and helps keep the beach in place and is often filled with activity and people sitting on its edge. The beach is restored and where once it was filled with riprap and rubble, now hundreds of people lie on blankets. On a sunny weekend day, the promenade is so full of people it is almost impossible to jog.

Further reading

Bays, J. 2002. "Principles and applications of wetland park creation," in *Handbook of water sensitive planning and design*, ed. R. L. France, 263–95. Boca Raton, FL: Lewis Publ.

Calloway, J. C., and J. B. Zedler. 2004. Restoration of urban salt marshes: Lessons from southern California. *Urban Environs.* 7:1055–85.

Collins, T. 2007. "Art, nature, health and aesthetics in the restoration of the post-industrial realm," in *Healing natures, repairing relationships: New perspectives in restoring ecological spaces and consciousness*, ed. R. L. France, 157–85. Winnipeg, Manitoba, Canada: Green Frigate Books.

France, R. L. 2003. *Deep immersion: The experience of water*. Green Frigate Books.

Gobster, P. H. 2007. Urban park restoration and the "museumification" of nature. *Nature and Cities*: 2:95–114.

Kidner, D. W. 2007. "Nature's memory: Restoration and the triumph of the cognitive," in *Healing natures, repairing relationships: New perspectives in restoring ecological spaces and consciousness*, ed. R. L. France, 69–93. Winnipeg, Manitoba, Canada: Green Frigate Books.

Kirkwood, N. 2004. *Weathering and durability in landscape architecture: Fundamentals, practices, and case studies*. New York: Wiley.

Krinke, R. 2001. "Overview: Design practice and manufactured sites," in *Manufactured sites: Rethinking the post-industrial landscape*, ed. N. Kirkwood, 125–49. London: Spon Press.

Mozingo, L. 1997. The aesthetics of ecological design: Seeing science as culture. *Landscape J.* 16:46–59.

Mozingo, L. 2007. "Constructing restoration ecologies: Nature, aesthetics, sites and systems," in *Healing natures, repairing relationships: New perspectives in restoring ecological spaces and consciousness*, ed. R. L. France, 185–98. Winnipeg, Manitoba, Canada: Green Frigate Books.

Reider, K. 2001. "Crissy Field: Tidal marsh restoration and form," in *Manufactured sites: Rethinking the post-industrial landscape*, N. Kirkwood, ed., 192–204. London: Spon Press.

Ryan, R. 2007. "Understanding the role of environmental designers in environmental restoration and remediation," in *Healing natures, repairing relationships: New perspectives in restoring ecological spaces and consciousness*, R. L. France, ed., 199–219. Winnipeg, Manitoba, Canada: Green Frigate Books.

chapter 3

California dreamin'—a reality
Multidimensional resource utilization in Arcata*

The story of the Arcata Marsh Wildlife Sanctuary is one of multidimensional resource utilization and the opportunities afforded when space, regardless of its context, is valued and imaginative actions taken to capitalize on those opportunities. The story also shows how a 30-year period of fruitful collaboration among many individuals and forward-thinking agencies can produce a pioneering and award-winning project.

The northern California city of Arcata in the 1890s was a small community situated on Humboldt Bay that was visited by three-masted schooners unloading mining supplies on a long pier that stuck out into the bay. The bay was characterized by a large saltwater marsh and extensive tidal mudflats, which had been used by the regional Amerindians for purposes of food and ceremony for thousands of years. The growth of the city had much to do with edges, spaces, and orientation to the sea. The bay edge was transformed by building a railway and then a highway upon the diked saltmarshes, which in turn were drained for industrial development and agriculture. Industrial activity consisted of a timber processing mill, a burner to dispose of sawdust in the days when it was legal to do so, a stacked lumber yard, a log pond, and a city corporation yard. The result was that Arcata had lost its orientation to the bay—its "bayview window" became closed. A big emphasis of the Arcata landscape regeneration project was therefore focused on interventions to help open up that bayview window.

Examination of early aerial photographs clearly shows the effects of diking and draining the wetlands. Inland from where the rail and road have cut the city off from ocean, all the saltmarshes have disappeared. Before the Coastal Zone Act was passed, industrial activities, including an open dump, had expanded into the drained marshes, and a series of oxidation ponds for wastewater treatment were built out into the bay. The famous mudflats, responsible for over 60 percent of all the oysters grown in California, were severely compromised as a result.

In addition to concerns about the shellfishery, other important issues were related to the health of the fishery for anadromous finfish, whose

* Adapted from a presentation by Robert Gearheart.

19

abundance in the state has been severely reduced through land management alterations. The city of Arcata has been involved in wastewater aquaculture: the raising of salmon in wastewater and their release to ocean for 4 to 5 years before their return to the effluent to which they were chemically imprinted. Other projects initiated by the city included urban stream restoration and examination of Humboldt Bay for other beneficial uses. It is for the pioneering use of artificial wetlands constructed for wastewater treatment that Arcata is best known around the world today. However, though the treatment wetland system was the initiation for the regeneration of Arcata's bayview window, many other developments occurred along the way, including pilot projects for phytoremediation research, and wetland habitat enhancement and creation for environmental education and public use.

Arcata, a city of 20,000 people in a county of about 130,000, developed a broad vision about how to regenerate its coastal waterfront. A close working relationship with elected city officials and staff members was essential to bring about projects, which were developed through a natural evolution of 25 years and not through a detailed, *a priori* master plan. That said, the city and its advisors operated under a set of general guidelines:

- Do not waste wastewater as it has value (as reflected in the city's motto "flush with pride").
- Support wastewater aquaculture and urban stream restoration.
- Maintain green belts and open spaces.
- Employ natural systems in infrastructure as much as possible.
- Encourage outreach extension of Humboldt State University in terms of student and faculty involvement in governmental and local issues (in other words, develop the regenerated landscape as a living laboratory).

The cumulative effects of shoreline development had resulted in decreased catches of oysters, clams, and fish, lower abundances of water birds, loss of salt and brackish water wetlands, loss of anadromous spawning sites in streams, and loss of natural drainage patterns with ditching and culverting. Arcata undertook a major program to attempt to turn around these deleterious environmental trends through the reclamation of industrial sites, urban drainage networks, agricultural land, landfills, and bay dredging. There was no conscious attempt to return the ecosystem to its predisturbance condition. For example, for agricultural land reclamation one of the problems was that the land had subsided to such a degree that once the system was opened up to the bay, it became impossible to restore the original type of wetland unless, in a few places, fill was brought in to build up a base upon which to create a semblance of the saltmarsh that had once existed there. Instead, a series of new reclaimed systems were constructed. Freshwater wetlands were created from the reuse of wastewater effluents and groundwater, which resulted in wildlife habitat at the same time as wastewater was being treated to a higher quality prior to its discharge. The natural structure and function of streams was restored. Dikes were removed to restore natural tidal flows to the saltwater wetlands, and tidal gates were installed to let fresh- and saltwater intermingle in the restored estuary. Aquaculture of oysters was developed based on using the nutrients in wastewater in order to grow such protein. The old landfill was buried under 2 to 3 feet of bay mud extracted from an adjacent bermed area inside which derelict agricultural and industrial lands were converted to enhancement wetlands. A major drainage ditch was restored into functional meandering estuary slough stream, and an industrial log pond was transformed into a new fresh-water marsh. All these efforts took place around the creation of a set of ponds for primary and secondary wastewater treatment and a series of treatment marshes for final wastewater "polishing."

The entire area making up the Arcata Marsh Wildlife Sanctuary includes 150 acres of open space for public access and passive recreation, in addition to the 30 acres of wildlife sanctuary and wastewater treat- ment enhancement wetlands. An interpretive center was built to focus on wetland awareness, research, and education. Friends of Arcata Marsh (FOAM), a nonprofit support group, was formed with a focus on inter-pretation activities, docent training, as well as, interestingly, advising the city on what it should do with the wetlands. The wastewater treatment facility is open 24 hours a day, 365 days a year, the strategy being that more access means more visitors, which means there are more uses for the site, and therefore more visibility and consequent support funding. The site includes 5 miles of trails for the public, an estuarine lake (Klopp Lake), and a created wetland fed with well water (Butcher Slough Marsh). A closed landfill was transformed into Mount Trashmore, which, as the highest location on the site, provides an overview of the entire area. The former landfill is rented from the city for one dollar a year with responsi-bilities for safety, but managers knew before purchase that there were no significant leachate problems. Nevertheless, the site is still monitored con-tinuously. Piers in front of the interpretive center, which are the remnants of the old plywood mill and burner, were left in place as a metaphor for the transition in transforming an industrial site into a nature sanctuary. The interpretive center itself is actually built on top of foundations for a large bucksaw, which had been used to saw apart large pieces of wood pulled out of the nearby log pond.

As ecology is all about edges, margins, and interfaces, these were maximized when creating and expanding the site's aquatic systems. Inside bends of meandering streams were leveled in order to provide conditions suitable for growing pickerelweed cord-grass in the saltmarsh restoration.

Over 200 species of birds have been sighted at the Arcata Marsh Wildlife Sanctuary (out of 250 species that are regularly found on the north coast of California), and the Audubon Society conducts weekly outings. Bird watching is the most popular recreational activity with many blinds and other observation structures located throughout the site. Fifteen species of mammals are routinely sighted, and 5 fish species are present, as well as a wide variety of aquatic insects. The large lake is stocked with steelhead trout raised from the wastewater aquaculture project for recreational fishing.

Trails wander around the entire site, including even the wastewater oxidation ponds. Public access to view and experience nature is the major organizing directive. The postindustrial site had actually reverted back to a seminatural state once it had been deemed of being no value. The city of Arcata was granted permission to obtain and use the site with the single proviso being that the public must be allowed access. An extensive monitoring program exists to obtain data about birds, people, traffic, and educational activities. Between 150,000 to 180,000 people a year visit the site, some taking docent-led tours organized by FOAM (Friends of Arcata Marsh).

The site has also become popular for many cultural events that on the surface have nothing to do with nature, per se, but are simply attracted to a scenic place. Such events include fundraising walks, weddings, family reunions, memorial services, and hospice activities. Significantly, the site has really become just as much a human as a wildlife sanctuary. Therefore, in addition to the seven thousand people a year that utilize the interpretive center (including over fifty school trips per annum), the overall popularity of the site provides an ancillary benefit in terms of introducing people who might not otherwise be the type to go out of their way to visit a nature center to issues of environmental awareness and education.

The Arcata story has drawn the attention of the worldwide community of engineers, biologists, and city managers, and the site has been used for many international workshops and training programs run by both the university and the city. Significantly, collection of data from this site has gone a long way to dispel the myth that a wastewater facility cannot be a wildlife sanctuary. Research conducted here has shown, for example, that the site is ten times more productive per unit area compared with other wildlife sanctuaries in the region. In the end, though, the city had lost an estimated 700 acres of wetlands over time, and the project has created 30 fully functional acres to counter the environmental deterioration produced by development.

One important lesson that emerges from the Arcata story is that a piece of ground with an intended function of treating wastewater to a high environmental standard can also become extremely valuable through the development of other layered uses, such as passive recreation park activities, environmental education, wetland habitat, carbon sequestration, increased property values (an adjacent new condo is called Marsh

Commons), public relations, and community marketing. The values accrued from investment in wastewater treatment has therefore produced multiple benefits:

- Alternative advance wastewater treatment (nitrification, denitrification, filtration) = $240,000 per year
- Recreation/public use (180,000 people × 1 hr x $6.50/hr) = $1,170,000
- Environmental education (10,000 people × 2 hr × $28/hr) = $560,000
- Wetland habitat (90 acres × $20,500/ac for mitigation) = $147,000
- Public relations (52 wk ×$750/wk) = $39,000
- Total annual community value = $2,156,000

The important take-home message is the relatively high values recognized for public recreation, and the bottom line is that the 30 acres that the city paid less than $100,000 for in 1983 now has an annual value of around $2 million.

The annual operations and maintenance costs are as follows:

- Personnel (2.0 full-time equivalents) – operational = $120,000
- Personnel (0.25 FTE) – administrative = $20,000
- Energy (150,000 kWh/yr) @ 0.08 kWh = $12,000
- Vegetation control = $5,000
- Monitoring = $10,000
- Equipment/parts – pump station = $2,000
- Total = $170,000
- Cost-million gallons (MG) @ 730 MG/yr = $232/MG/yr

This amounts to an annual benefit/cost ratio of 12.8, which given that anything more than 1 is better than most waste treatment plants, is a phenomenal return on the investment. Once other benefits are overlaid, the true value of land becomes obvious. The operational lesson is the necessity of sitting Parks and Recreation Department personnel around the planning table in addition to Department of Public Works officials, in order for the former to lay out the multiple benefits ensuing from nontraditional wastewater management in order to counter the argument that such approaches are too expensive or that there is not enough valuable space to "sacrifice" to such a project.

Reclaimed wastewater and harvested stormwater are the most available and reliable sources of water for maintaining freshwater wetlands in urban settings. It behooves us, therefore, to find ways to take advantage of these two resources. Such treatment and reuse systems create space for passive recreation opportunities on damaged or derelict postindustrial and postagricultural land. The economic and cultural environments of many municipalities are ripe for development of such community-based, integrative, innovative, and sustainable projects that arise just as much from bottom-up as they do from top-down processes. A key element is the need for participatory research, wherein stakeholders establish needs, goals, and assist in identifying resources, collecting data, constructing, planting, monitoring, managing, and operating these reclaimed areas.

The Arcata story illustrates an important paradigm shift that is beginning to develop in wastewater management, as shown in the table below.

Technical Concerns	
Present	Future
Standard-technology	Performance
Equipment sophisticated	Appropriate
Capitalization	Sustainable
Operations complex	Simple
Nonwetland/agricultural	Wetland/reclaimed landscapes

Societal/Economic	
Present	Future
Healthy individual	Healthy community
Centralization	Decentralization
Technology-based	Community-based
Societal cost	Community opportunity
Subsidized	Self-supporting

Resource/Environment	
Present	Future
Resource protection	Resource enhancement
Energy intense	Low energy needs
Discharge/disposal	Reuse/reclamation
Technological approach	Ecological approach

The current paradigm minimizes innovation, utilizes equipment with a limited lifespan, impedes the ability for the users to decide what to do, and provides false economies of scale. In contrast, the future paradigm is based on taking wastewater and using it one step further in terms of not just protecting beneficial uses but actually *producing* new beneficial uses. For this shift to occur, it is beneficial for technical individuals to remain on the outside as information sources and critics and to help in the translation and education processes but not necessarily in the identification of the alternatives, which are better generated by the public. These technical advisors serve their roles most usefully if they possess a good grasp of fundamentals in terms of understanding the science behind the regional or local hydrology, climate, and natural resources.

Further reading

Austin, D. 2010. "Advanced ecotechnology for decentralized wastewater treatment and reuse," in *Restorative redevelopment of devastated ecocultural landscapes,* ed. R. L. France, 391–412. Boca Raton, FL: CRC Press.

Bays, J. 2002. "Principles and applications of wetland park creation," in *Handbook of water sensitive planning and design,* ed. R. L. France, 263–95. Boca Raton, FL: Lewis Publ.

Campbell, C. S., and M. H. Ogden. 1999. *Constructed wetlands in the sustainable landscape.* New York: Wiley.

Environmental Protection Agency. 2000. *Manual: Constructed wetland treatment of municipal wastewater.* Washington, DC: EPA.

France, R. L. 2004. *Wetland design: Principles and practices for landscape architects and land-use planners.* New York: W.W. Norton.

France, R. L. 2010. "Linking water treatment with wetland restoration: Engineering challenges and associated benefits," in *Restorative redevelopment of devastated ecocultural landscapes,* ed. R. L. France, 375–90. Boca Raton, FL: CRC Press.

Kadlec, R. H., and S. Wallace, 2007. *Treatment wetlands.* Boca Raton, FL: CRC Press.

Kadlecik, L., and M. Wilson. 2007. "Cultural and environmental restoration design in northern California Indian country," in *Handbook of regenerative landscape design,* ed. R. L. France, 315–43. Boca Raton, FL: CRC Press.

Rasmussen, M., and S. Hurley. 2007. "Coastal ecosystem restoration through green infrastructure: A decade of success in reviving shellfish beds with a storm-water wetland," in *Handbook of regenerative landscape design,* R. L. France, ed., 161–78. Boca Raton, FL: CRC Press.

Saunders, W. S. (Ed.) 2007. *Nature, landscape, and building for sustainability.* Minneapolis: Univ. Minn. Press.

Van der Ryn, S., and S. Cowan. 2007. *Ecological design.* Washington, DC: Island Press.

Wallace, S. 2010. "Infrastructure framework for decentralized wastewater planning," in *Restorative redevelopment of devastated ecocultural landscapes,* ed. R. L. France, 413–26. Boca Raton, FL: CRC Press.

chapter 4

California case study lessons

- Successful regenerative landscape design projects for recreation:
  - Can capture the imagination and emotions of the public
  - Need to be conducted in concordance to strict scientific and engineering criteria
  - Are part of green infrastructure and can solve existing technical or management-related problems
  - Provide value-added community benefits in the creation of popular public space
  - Sometimes are in opposition to the traditional mindset of engineers, who nevertheless have to be brought on board for eventual success
  - Often require an important role for landscape architects in defining borders in dense urban settings
  - May sometimes be accepted only once the public is taken out to visit the site and have explained to them the physicality of the proposed designs
  - Need to factor in a period of generating conceptual designs for public review and discussion, especially if proposing innovative concepts.
- Public regenerative landscape design projects are complex and may take half a decade or longer to be implemented.
- For severely altered urban landscapes, it is often impossible to return to predevelopment conditions.
- Opposition to regenerative landscape design can come from unlikely sources, such as your imagined allies; for example, engineers may often be better supporters than parks officials or environmental activist groups.
- Complex developmental site histories mean that such landscapes are palimpsests, which can also include unrecorded interventions by Amerindian peoples.
- Recreational concerns are extremely important in urban settings because of the absence of open-space, and this can be a major challenge to address and accommodate because of divergent wishes of different activity or interest groups.

- Freed of a need to ascribe to an impossibility of setting the clock back in time in terms of historical fidelity, regenerative landscape designers can undertake innovative designs without the need to disguise human artistry.
- A team of qualified technical scientists and engineers is de rigueur for rebuilding ecology from the ground up in severely damaged or altered sites.
- Functional ecology may be difficult to recreate because of space constraints and programmatic limitations.
- Regenerated urban landscapes can be extremely popular civic spaces.
- Regeneration landscape design for recreation and ecotourism can be combined with local economy initiatives toward creating a new green economy.
- It is possible and indeed desirable to piggyback landscape regeneration with infrastructure improvements, moving design from simply protecting old, to creating new, beneficial uses through a program of integrated land reclamation.
- In highly disturbed urban situations it may be futile to attempt to return to predevelopment conditions.
- One valuable strategy is to do everything possible to facilitate the public access and recreational use of regenerated landscapes in order to boost site visibility and raise project funding.
- It is important to retain remnants of previous industrial uses on site in order to celebrate rather than to overwrite the past.
- Nature observation and ecotourism are major drawing cards, and facilitating such recreation may ultimately be as important as nature re-creation.
- Because development of layered, alternative uses, such as programmed cultural activities, greatly increases the value of regenerative design projects, it is beneficial to use parks and recreation department personnel in initial planning.
- Economic analyses are important in the interpretation of the many benefits accruing from landscape regeneration.
- Programs of participatory research through use of knowledgeable technical advisors is important for local education and for generating widespread interest in certain innovative projects.

Recovery processes and design practices for regenerating derelict landscapes for recreation and ecotourism

chapter 5

From the Las Vegas Wash to Clark County Wetlands Park*

Background and early project development

There are few less likely places for healing the split between humans and nature than Las Vegas, Nevada. Perhaps nowhere embraces, if not outright celebrates, excess with such proud zeal as does this desert metropolis. More than the glitzy neon splendor, the tacky tourist "amenities," or the nonstop gambling dens of imagined (or possibly realized) sin, is the proud defiance with which this desert city alternately flaunts or ignores its water. As one article in a national magazine stated, "Las Vegas is America's city of fantasy, and water, not wealth, is its greatest fantasy of all...displayed more lasciviously than sex" (Figure 5.1). On the one hand, tourists along the Strip (Las Vegas Boulevard) are exposed to water displayed as grandiose, carnival-like features, such as imagined Italian fountains, Venetian canals, tropical lagoons, or pirate-infested Caribbean seas. On the other hand, how Las Vegas is dealing with restoring the environmental degradation caused by its stormwater runoff is a story worthy of celebration. The Clark County Wetlands Park System is the largest open space (3,000 acres) in the urban Las Vegas Valley and is considered the hub of the regional trail system. Its creation, generated through development of an incredibly comprehensive master plan, represents over 20 years of involvement and planning.

The Park is a place in the hot and arid desert that is very special, being rich in biological, geological, and cultural resources. Although located only 15 minutes off the world famous Las Vegas Strip, the Clark County Wetlands Park couldn't be more different. The Park is situated in the Las Vegas Wash, which was once a modest grassy meadow area that has over time been transformed (primarily as a result of urban runoff and treated wastewater discharge) into an extensive wetland. The area is the lowest point in the Las Vegas Valley and is surrounded by mountains ranging in elevation from 7,000 to 11,000 feet. All the surface water and rainwater falling within the 1,500-square mile hydrographic

---

* Adapted from a presentation by Jeff Harris, Mark Raming, Vicki Scharnhorst, and Becky Zimmerman.

Figure 5.1 Greening of a desert suburb of Las Vegas.

basin finds its way through the Las Vegas Wash and the Wetlands Park to where it eventually feeds into Lake Mead and then the Colorado River (Figure 5.2). The Wash contributes about 2 percent of the water flowing into Lake Mead. The Wetlands Park is located in the southeast portion of the Las Vegas Valley and is fed by inflows that are ephemeral washes. Some areas, for example the Rainbow Gardens, are very significant in terms of their geology. As the area is administered by the United States Land Management Bureau of Reclamation, the National Parks Service and local municipalities, development of a cooperative management plan was essential.

The first human visitors to the Las Vegas area were Amerindians who, upon crossing the hot Mojave Desert, found an oasis of artesian springs. Forty-one historic sites have been identified (rock shelters, cleared circles, artifacts, trails, etc.), some dating from possibly as early as 7,000 B.C.E. For centuries, the Paiute and Pione tribes lived along the verdant wet meadows, which provided them shelter, food, and water, and then upon "discovery" in the early-nineteenth century by Spanish explorers, the location became

Figure 5.1 (continued)

a critical stopover along the old Spanish trail from Santa Fe (New Mexico) to Los Angeles. The frontiersmen John Freemont and Kit Carson visited the meadows area in the 1850s while establishing a trade route from Salt Lake City (Utah) to California. By the late-nineteenth century, Mormon settlers had also found the valley, now referred to by its Spanish name for meadows: "Las Vegas." These settlers built a fort, Las Vegas Mission, to protect immigrants and the U.S. mail from Amerindian attacks; they also established farms and ranches and began diverting the abundant water for irrigation and everyday use.

Farms and ranches continued to spring up in the first decades of the twentieth century, and manganese was mined on and off between 1917 and 1961. Up until the 1930s, Las Vegas was a small railroad town with a population of about 5,000. As the city grew, the Wash served as a natural channel for discharging the treated wastewater; as a result, the system began to change. By 1955 the Wash, which before then had been an ephemeral stream, began to flow year-round because of the discharge of treated wastewater effluent. These nutrient-rich discharges soon transformed the

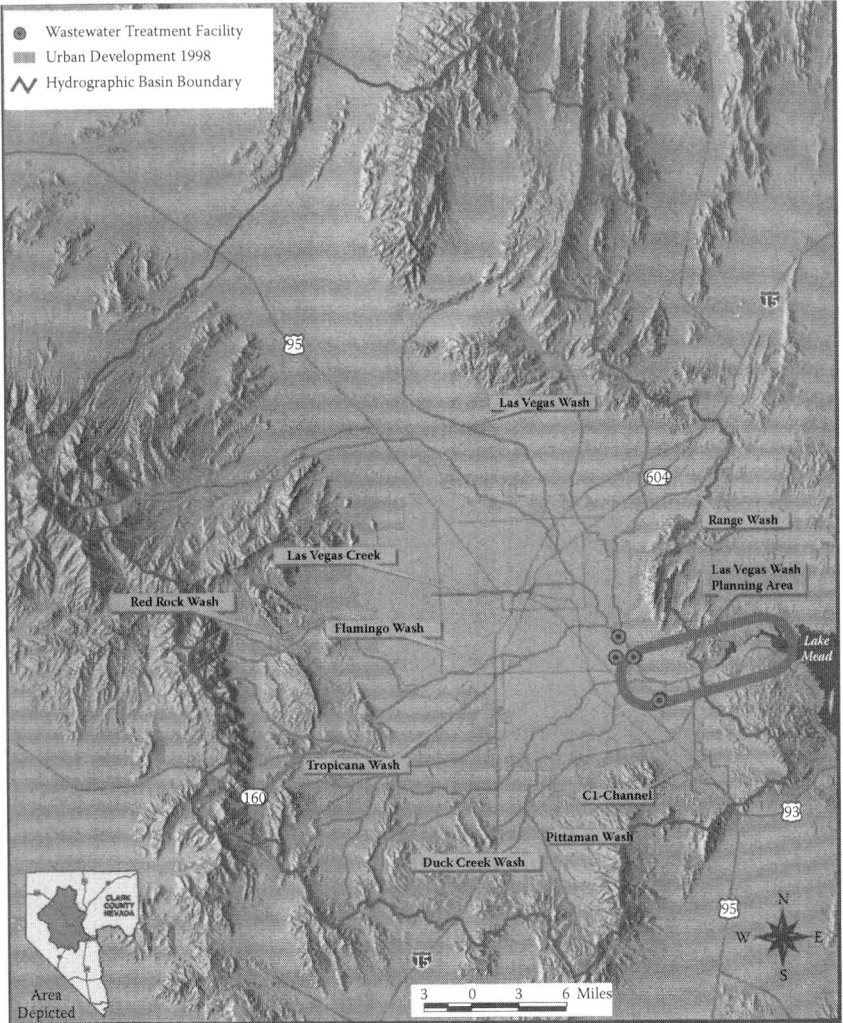

Figure 5.2 Location of the Las Vegas Wash in the Las Vegas Valley.

wet meadows into a lush, cattail-dominated wetland that began to attract thousands of birds (Figure 5.3). By the late 1960s, the system was out of balance, the wetland vegetation unable to absorb the increased runoff resulting from urbanization, and the "excess" water beginning to erode the fine-grained soils of the fragile floodplain. Then in the 1970s, a large fire destroyed vast expanses of the vegetation, exposing the soils and leading to erosion, which in turn contributed to the channelization of the Wash and destroyed lateral wetlands. Further, by concentrating the stormwater flows into the increasingly deeply cut channel (Figure 5.4), the riparian vegetation

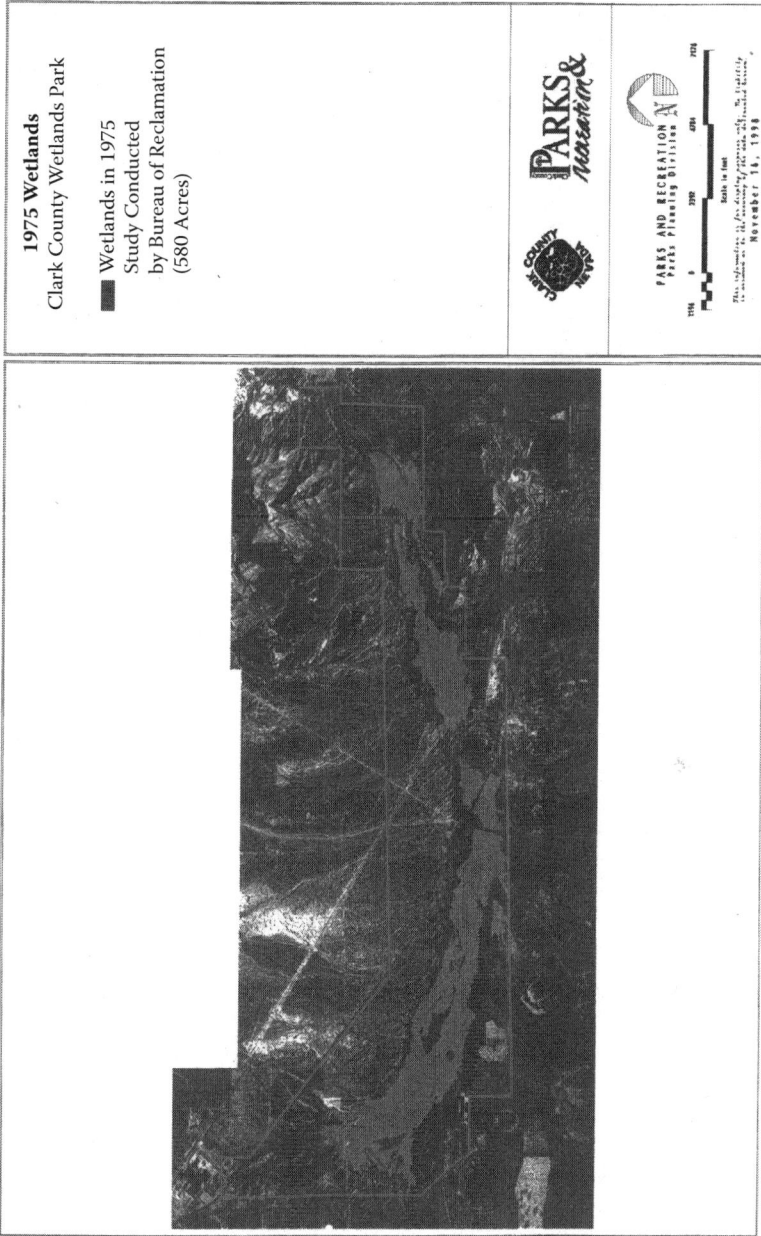

Figure 5.3 Extent and typology of verdant wetlands of the Las Vegas Wash in the 1970s.

Figure 5.3 (continued)

was left high and dry, such that more and more wetlands disappeared. The rate of wetland destruction and loss was staggering—an 80 percent reduction in surface area from 1975 to 1982. Thus, the natural purifying role played by these ecosystems no longer occurred, and the water flowed into Lake Mead without the benefit of this final phytoremedial "polishing."

Figure 5.3 (continued)

Figure 5.4 Severe channel cutting and erosion in the Wash caused by storm events.

Figure 5.4 (continued)

Interestingly, because flows in the Wash increased proportionally with population growth, it was the very process which had initially created the wetlands that began to contribute to their ultimate undoing

Over time, the Wash was looked upon by most residents of Las Vegas with derision, if regarded at all. An abused and marginalized land-scape, the Wash became the regional dumping ground for trash and was populated by all-terrain vehicles, transients, and locals discharging their firearms (Figure 5.5).

The current Wetlands Park in Las Vegas is living documentation of all of these antecedent occupations, whose traces in the form of trails can be found in the Park, bearing witness to the hardships of those early

Figure 5.4 (continued)

Figure 5.4 (continued)

Figure 5.5 Misuse of the current Las Vegas Wash before Park establishment.

years. But why is the site so important to modern Nevada? First, the abundant meadows provide the water and habitat needed to maintain a wide variety of desert animal life. Second, having wetlands beside the urban center provides the opportunity for an alternative type of recreation and environmental education that cannot be duplicated anywhere else in the surrounding Mojave Desert.

Figure 5.5 (continued)

The concept for the Park began over 30 years ago by a group of citizens with minimal government support. Although largely composed of a small group of birders with binoculars, several professional biologists were present who envisioned creation of a "wetlands park." One of these individuals, Vern Bostick, who continued to visit the site into his nineties, was instrumental in keeping the dream alive. During these early years, this citizen group was very effective in identifying and working to limit peripheral encroachment of developments into the area, at the same time as serving as the general stewards of the site. In the early 1990s, the first significant public funding became available; after that, federal agencies became interested in the concept of a wetlands park. It is in surmounting the challenges to implementing that concept where the Clark County Wetlands Park project offers valuable lessons. The master plan that was developed to guide the restoration of the Las Vegas Wash into the Clark County Wetlands Park is a signature document about how to undertake and accomplish the successful reclamation of degraded, high-visibility sites.

Desert washes are by definition very dynamic. The current floodplain (Figure 5.6) is wide and is composed of sands, silt, and gravel of a high erodibility because of its original nature as a floodplain formed 24 million years ago. Two catastrophic storms in 1978 and 1984, for example, caused severe erosion of the stream channel. Getting control of this extremely dynamic environmental system would not be an easy task. First, the soils must be

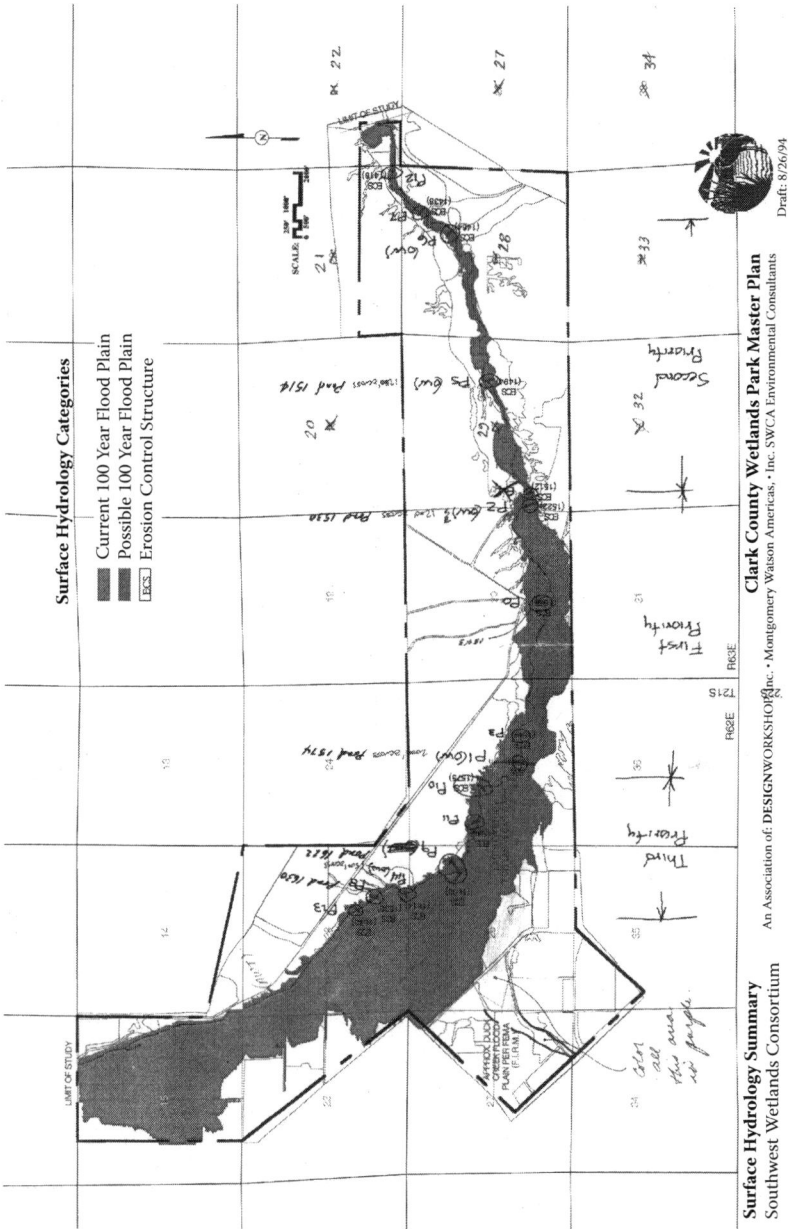

Figure 5.6 The 100-year floodplain in the Clark County Wetlands Park.

stabilized in order to provide an adequate foundation for constructing facilities in the Wash over its course from the Northshore Bridge on the outskirts of Las Vegas to the edge of the Lake Mead National Recreation Area. With up to a 30-foot headcut of erosion resulting from a single storm event, the surging water can carry away hundreds of thousands of tons of sediment, abandoned cars and shopping carts, and anything else that was in the Wash right into Lake Mead. After such a flooding event, the natural vegetation is stripped away, leaving the area susceptible for colonization by weeds and other invasive species which typically have little wildlife value. The physical extent of the wetlands has decreased considerably. In 1985, there were over 2,000 acres of marsh and riparian habitat, whereas less than 700 acres remain today. Erosion is by no means the only problem; illegal dumping and vandalism continue to challenge the development of the Park as well. Sometimes, however, the graffiti can provide a valuable service as a striking reminder of environmental change. For example, a recent photograph of the Northshore Bridge showed graffiti 40-feet in the air at an elevation where the ground surface had been in 1978 prior to erosive scouring. Today, the ever-increasing amount of runoff and consequent erosion threatens the stability of this bridge, which may have to be closed for transportation safety reasons given that it is routinely traversed by vehicles at speeds around 50 miles per hour.

When the project was started, much of the area was a patchwork of land ownership (250 individual owners) that complicated management decision making and park development (Figure 5.7).

Located immediately to the south of the Clark County Wetlands Park is the City of Henderson's Water Reclamation Facility and Bird Viewing Preserve. This wastewater-waterfowl preserve has become a major birding area supported by a local chapter of the Audubon Society and offers a striking example of how treatment wetlands can become the central foci in landscape reclamation and public visitation (see Appendix).

The Clark County Wetlands Park Master Plan established four goals to help guide development activities:

1. Develop recreational and tourism opportunities, based on public needs, that are compatible with the conservation and restoration of the Las Vegas Wash. Specifically, (a) create passive park and recreation facilities for public enjoyment; and (b) develop recreational opportunities in the context of the overall recreational system in Clark County.
2. Create social benefits for the Las Vegas Valley by providing opportunities for area residents to gain a sense of community pride and ownership of the Park. Specifically, (a) provide opportunities for community involvement and activism; and (b) receive input from adjacent and nearby property owners throughout the planning process.

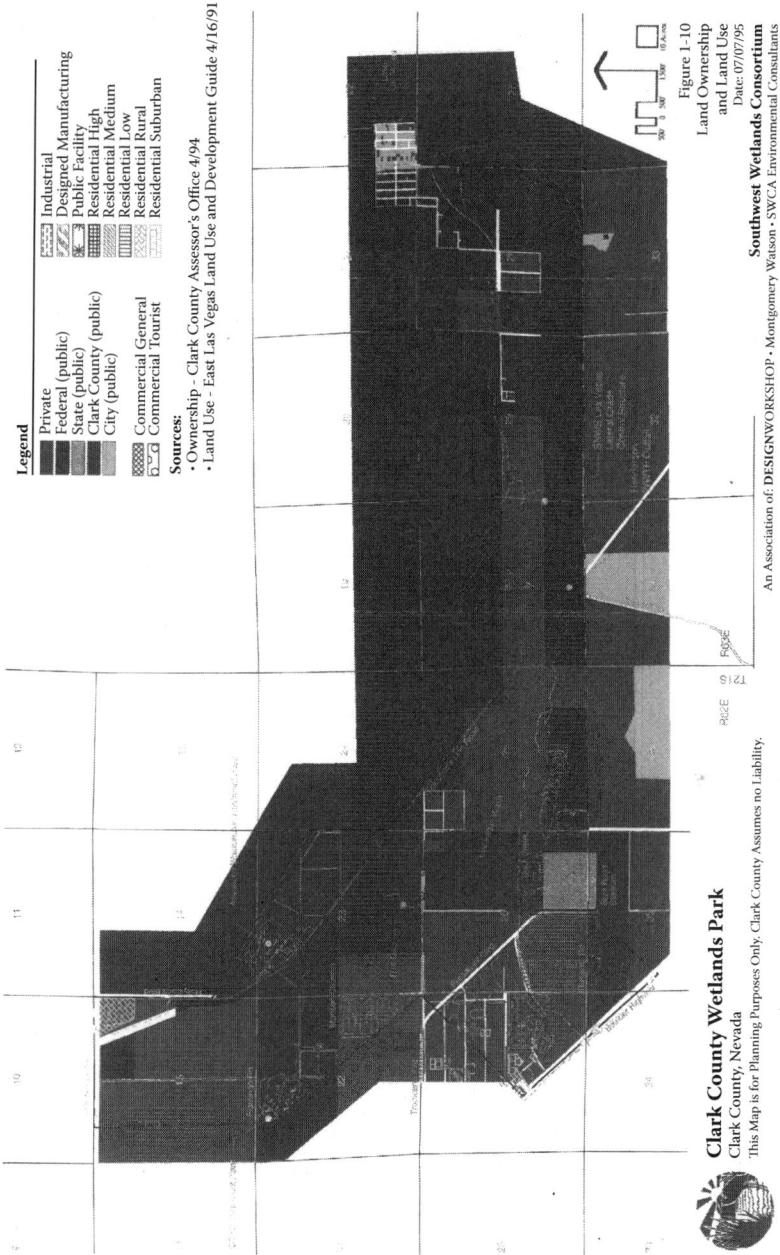

Figure 5.7 Land ownership and land use patterns in the Wetlands Park.

3. Create educational opportunities to convey the importance and significance of the Las Vegas Wash through various media. Specifically, develop interpretive opportunities to educate the public on human impacts on wetlands, the significance of wetlands, and of desert environments; and (b) create opportunities for hands-on education on the history and ecology of this unique environment.

4. Conserve and restore natural resources by protecting and enhancing the ecological resources of the Las Vegas Wash. Specifically, stabilize eroded areas and improve water quality; and (b) communicate the significance of these resources with an interpretive education program.

From the start, it was deemed essential to provide some context for planning the present Park by examining comparative projects in order to gain a sense of scale and understanding of challenges likely to be faced. Specifically, a handful of arid region parks were studied to determine levels of development, visitation, and management, in addition to interpretive facilities, environmental education outreach, resource preservation, habitat enhancement, and operational costs. From this cross-system analysis, a set of commonalities emerged as guiding principles for planning the future Clark County Wetlands Park:

- Select quality rather than quantity.
- Stay with one theme and avoid fragmentation.
- Give visitors a sense of scientific fact rather than environmental emotionalism.
- Stewardship and volunteerism are critical components.
- Address maintenance and law enforcement early in the planning process.

The overall goal for the Clark County Wetlands Park is to create an environment that has a timeless spirit and reflects the characteristics of its unique genesis and desert location. The Park, through its design, is intended to capture the grandeur and beauty of its setting and at the same time convey an historic sense of the region. Finally, it is hoped that following rehabilitation, the Park will function as an "urban wild" that sustains wildlife at the edge of the fastest-growing community in the country.

Integrated planning and site design process

In 1991, Nevada residents approved by ballot a bond initiative of $13 million to create a park in the Las Vegas Wash. Two years later, the Southwest Wetlands Consortium was formed (a partnership of individuals from Design Workshop Inc., Montgomery Watson Harza, and SWC Environmental

Consultants) to begin the master planning process. The guiding vision was to initiate a process whereby all those interested in the concept of the Park (agencies as well as individuals) could contribute to its creation.

The integrated planning and design process for the Las Vegas Wash needed to be comprehensive and compelling. Issues that had to be addressed included loss of wetlands, degraded site conditions, community perceptions, urban encroachment, and water quality. In particular, answers had to be found for the following concerns:

1. How to stop the wetlands loss and hopefully how to possibly gain back some area of previously lost wetlands.
2. How to turn the place from being a dumping ground into a recreational retreat.
3. How to rehabilitate the Wash so area residents can use it and take pride in the new regional park.
4. How to create a synergy of edges along the park-to-urban interface.
5. How to improve the water quality within the parameters of financial constraints.

Many consulting projects are compromised at the very start by having only a little time to accomplish a job. In contrast, this development team has had the luxury to have worked on the project for a decade, and the process is still ongoing. Generating the master plan itself was a 2-year process that combined innovative planning techniques and logical conclusions to create a vision that the pubic could buy into at the same time as providing the foundation for obtaining multiagency support. This latter was extremely important given that there are over thirty agencies and organizations involved in decision making for the Park. The primary elements for the master plan included public participation, inventory and analysis, programming, concept development, alternative concept development, preferred option selection, plan refinement, and documentation. It is important to note that simultaneous to the master plan development, the same team of consultants was preparing a programmatic environmental impact statement and environmental assessment for the erosion-control structures with the idea that if all these activities were happening simultaneously, the implementation timeframe could become more efficient and accelerated.

Public ownership of a master plan is imperative if it is to have any chance of succeeding. There was a need to focus the advocacy that had been built up for this land over the previous three decades, and a recognition that public input was essential to decide what type of rehabilitation should occur. As a result, the public was kept informed about the progress on the project through a series of newsletters (these were the days before widespread email use) and meetings. One of these meetings presented the analysis work and solicitations about the future of the Wetlands Park and asked about what particular types of uses the public wanted to see developed there. And another set of meetings provided an opportunity to review the preliminary Master Plan concepts.

Because effective planning depends on a good information base, a large part of the project was devoted to gathering and analyzing information. The site inventory included detailed information on geology (Figure 5.8), surface and subsurface hydrology (Figure 5.6), vegetation (Figure 5.9), visual resources (Figure 5.10), land use and ownership, existing infrastructure conditions, buildability, as well as delineating wetlands, documenting archeological sites, and conducting a survey of the desert tortoise. One lesson learned in this process was that the original schedule severely underestimated the amount of effort it would take to gather information that is current and comprehensive, as well as the time needed to verify the digital data by onsite, ground-truthing surveys.

To determine appropriate areas for park development activities and identify potential impacts of this development on biological and cultural resources, the study area was divided into six zones based on biogeochemical characteristics (Figure 5.11). Zone 1 is not suitable for interpretation and education because of minimal vegetation and low wildlife diversity (Figure 5.12). As a consequence, the area is appropriate for active and passive recreation. Zone 2 provides excellent opportunities for interpretation, education, and passive recreation because of its considerable wildlife diversity. Because zone 3 lies within the floodplain, it is not suitable for locating a major facility, although its wildlife diversity is high. The disturbed upland area in zone 4 makes it suitable as a location for active recreation. Zone 5 provides opportunities for interpretation, education, and both passive and active recreation.; because of steep slopes, the area is unsuitable for construction of major or minor facilities. Zone 6 provides opportunities for interpretation and education about cultural resources and endangered species habitat, and it is also appropriate for building facilities if the rare habitat areas are avoided.

Public workshops were important for guiding and refining the master plan scenarios. Lists of appropriate and inappropriate uses were derived with a set of prioritization exercises. Desirable uses that were identified included: environmental education for all ages, especially for youth; educational facilities, such as a museum or interpretive center serving both local residents and tourists; preservation, restoration, and conservation of wildlife and the habitats that support them; and trail amenities for activities such as walking, hiking, bird watching, and mountain biking. Undesirable uses for the Wetlands Park included automobile and noise-producing activities such as boating, model airplane racing, dirt biking, four wheeling, or all-terrain vehicles; subdivisions, casino

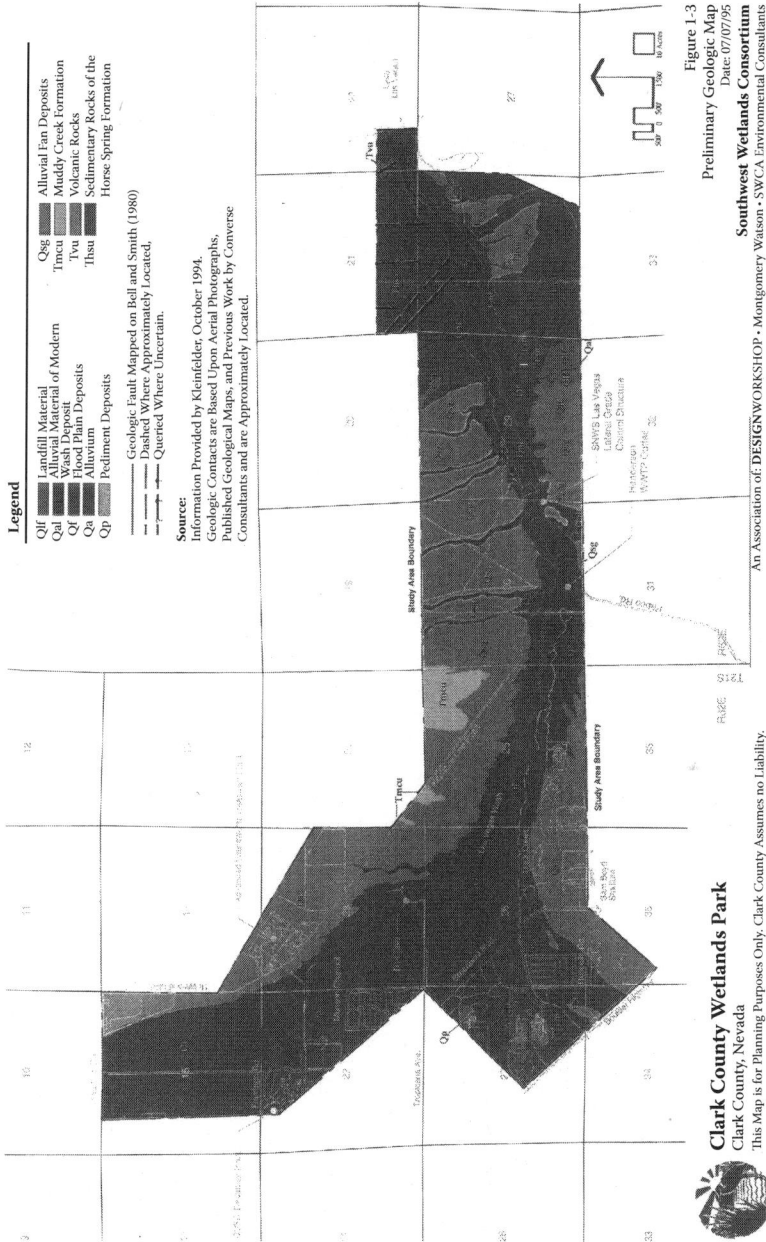

Figure 5.8 Site inventory of geology.

Figure 5.9 Site inventory of vegetation communities.

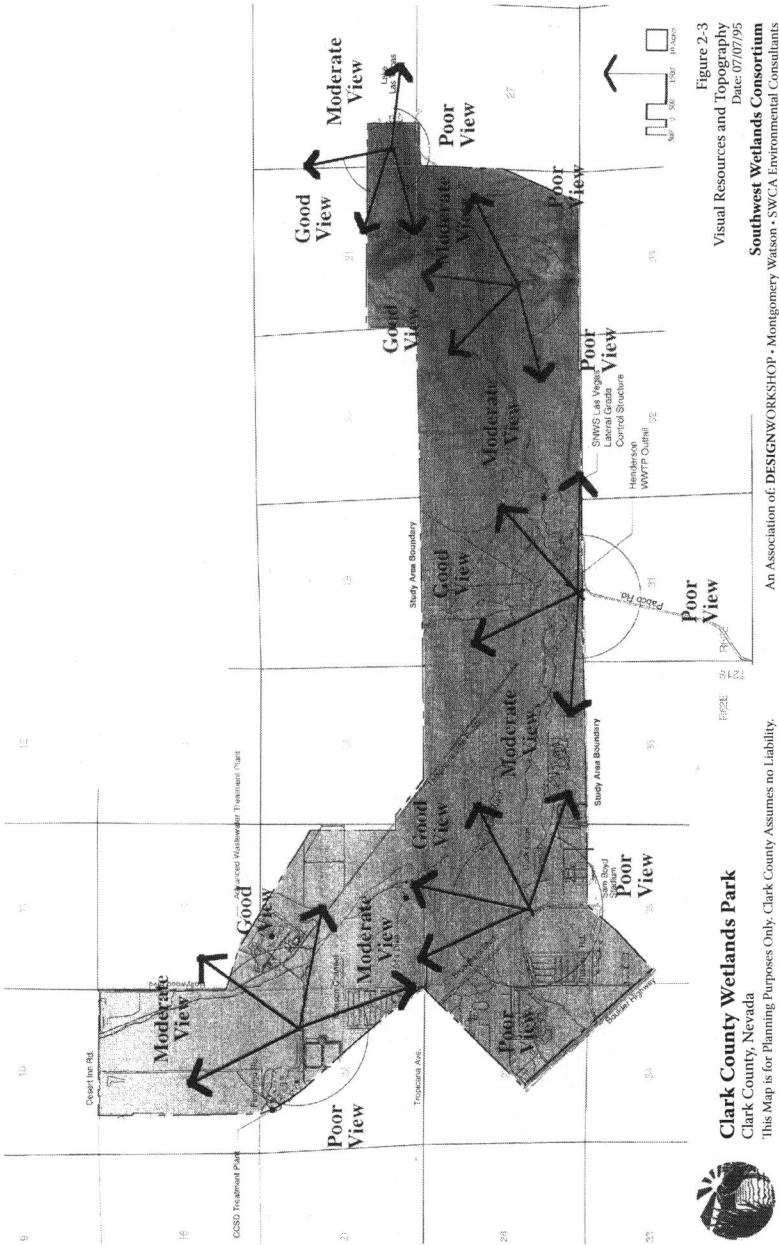

Figure 5.10 Site inventory of visual resources.

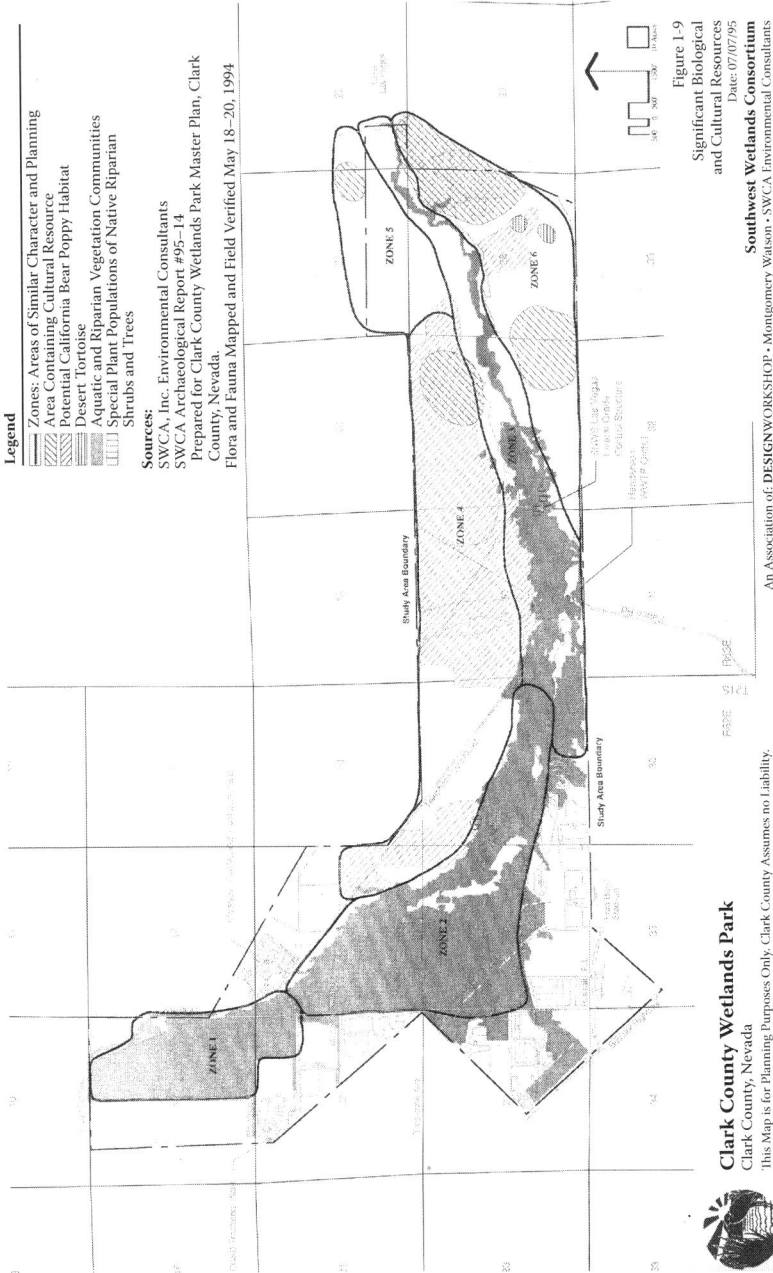

Figure 5.11 Division of the Wetlands Park into six planning zones of distinct characteristics.

ZONE	ACTIVITIES	RESOURCES	HABITAT ENHANCEMENT	EROSION CONTROL STRUCTURES
1	Active and Passive Reception; Major and Minor Facilities	Tamarisk community with limited habitat quality	Creating openings in tamarisk community in conjunction with facility construction	None
2	Excellent for Interpretation and Education and Passive Recreation; Active Recreation located outside wetlands	Diversity of wetland habitats including an area of alkali wetland; Confluence of Duck Creek and several other surface water sources; Small population of arrow weed located in a drainage north of the wash	Riparian and wetland habitat enhancement in areas of tamarisk and common reed along Wash channel and Duck Creek	Eight erosion control structures and repairing D14 dike. Potentially two structures and the D14 dike could be designed to pond water
3	Interpretation and Education; Passive Recreation; Active Recreation and Minor Facilities limited to upland edges of the Zone	Riparian habitat; Population of Fremont Cottonwood near the Las Vegas Valley water lateral; Severe Wash channel erosion and head-cutting downstream of Telephone Line Road	Riparian and wetland habitat enhancement near Wash channel in conjunction with proposed erosion control structures	Seven erosion control structures. Potentially, two structures could be designed to pond water
4	Interpretation and Education; Active and Passive Recreation; Major and Minor Facilities	Disturbed upland habitat from off-road vehicle use and trash dumping; Potential California bear poppy habitat in gypsum soils	Fencing out off-road vehicle traffic, removing trash, and restoring disturbed areas	None
5	Interpretation and Education; Active and Passive Recreation	Upland habitat; Archeological and historic sites; Upland habitat somewhat disturbed by off-road vehicle use	Fencing out off-road vehicle traffic, removing trash, and restoring disturbed areas	None
6	Interpretation and Education; Passive and Active Recreation; Major and Minor Facilities; Location of activities determined by the presence of significant resources	Upland habitat; Archeological and historic sites; Desert tortoise burrows; California bear poppy habitat; major disturbance by gravel mining operations	Fencing out off-road vehicle traffic, removing trash, and restoring disturbed areas; no habitat enhancement in gravel mining areas	None

Figure 5.12 Biological and cultural characteristics of the six planning zones.

development, or commercialization; concrete-lined flood channels; trash dumping; and target practice, shooting, and hunting.

Analyses of contextural opportunities, comparative facility research, and local needs assessment helped form the programming decisions. All this comprehensive information led to the development of three alternative themes that were developed into physical plans to enable the public to understand the landscape implications of each. The alternative plans varied in their general characteristics but shared common overall characteristics and specific remedies identified for countering the severe erosion that was threatening the site. These included cleanup and debris removal, fencing and security, severe and enforced fines for dumping, nature of erosion control structures, habitat enhancement and restoration goals and methods, prohibiting use of all-terrain vehicles and hunting and target practice, interpretive signage, and overflow parking. Elements included in each of the alternatives with differing levels of use included trails, public areas, trail heads, habitat enhancement, visitor facilities, interpretive exhibits and signage, road access, and the extent of created or enhanced wetlands behind one or more erosion-control structures.

Distinct characteristics were developed for each of the alternatives of "Conservation," "Recreation," and "Full Development" that helped in comparing and contrasting them. The Conservation Alternative (Figure 5.13) stated that the main priority was to conserve and restore the land with public use kept to a minimum; in other words, public use restricted to the perimeters and being concentrated at a visitor center, which was simply an information kiosk. The Recreation Alternative was based on maximizing the recreation opportunities of the entire landscape of 3,000 acres. And the Full Development Alternative even allowed for opportunities to undertake land trades, such as selling some land at the northern periphery to a developer in order to create a huge amount of revenue to be used in the Park, with facilities programmed for a maximum public interaction including a 30,000-square foot visitor center and many other features.

In the Integrated Alternative (Figure 5.14), the Wetlands Park would become an environmental and recreational resource that emphasizes habitat enhancement, recreational facilities, and interpretive features for the enjoyment and education of visitors. The goal of this alternative was to synthesize the recommendations from each of the others in a manner that appropriately balanced conservation, recreation, and revenue-generating objectives. Elements included in this alternative were wetland pond areas maintained behind raised erosion-control structures, establishment of more desirable riparian vegetation, a visitor center that offers interpretive exhibits and other displays, an environmental education camp, a research/education facility, a one-way scenic drive along the south side of the Wash, and a multiple-use trail along both sides of the Wash that includes links to trails that provide access to Rainbow Gardens and to the city.

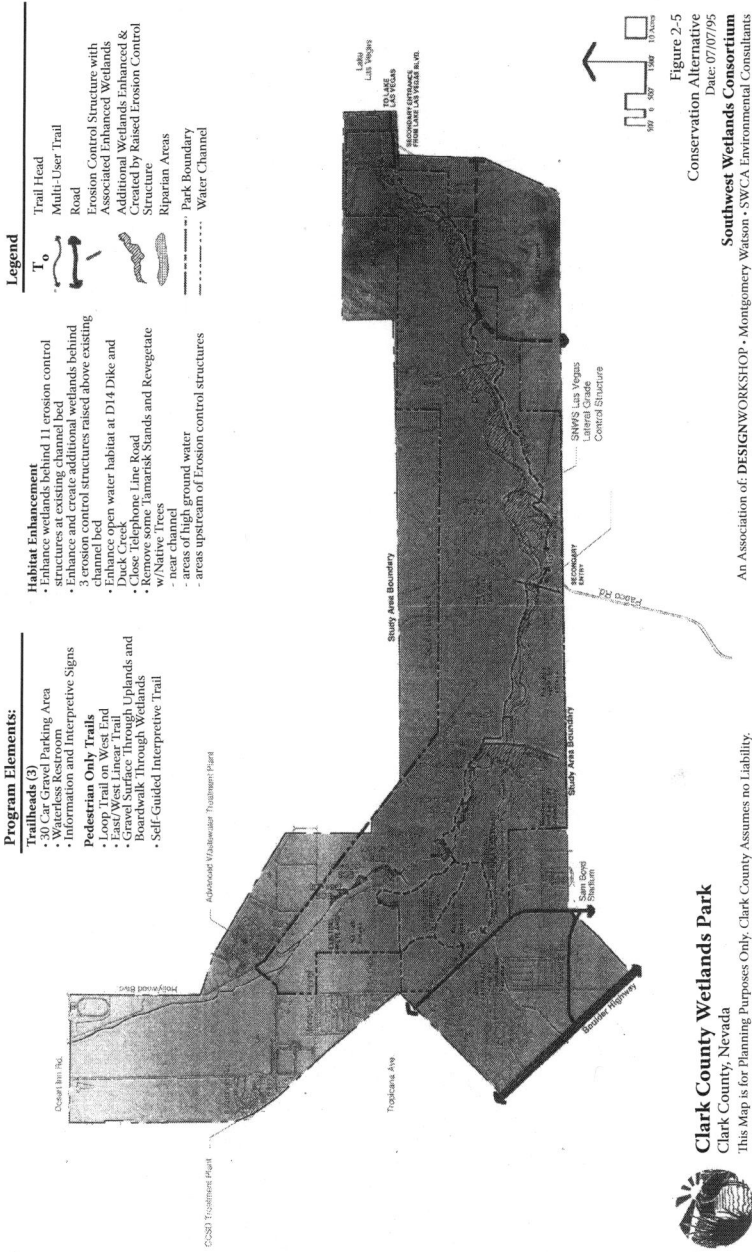

Program Elements:

Trailheads (3)
- 30 Car Gravel Parking Area
- Waterless Restroom
- Information and Interpretive Signs

Pedestrian Only Trails
- Loop Trail on West End
- East/West Linear Trail
- Gravel Surface Through Uplands and Boardwalk Through Wetlands
- Self-Guided Interpretive Trail

Habitat Enhancement
- Enhance wetlands behind 11 erosion control structures at existing channel bed
- Enhance and create additional wetlands behind 3 erosion control structures raised above existing channel bed
- Enhance open water habitat at D14 Dike and Duck Creek
- Close Telephone Line Road
- Remove some Tamarisk Stands and Revegetate w/Native Trees
  - near channel
  - areas of high ground water
  - areas upstream of Erosion control structures

Legend

T₀ — Trail Head
— Multi-User Trail
— Road
— Erosion Control Structure with Associated Enhanced Wetlands
— Additional Wetlands Enhanced & Created by Raised Erosion Control Structure
— Riparian Areas
---- Park Boundary
······ Water Channel

Figure 2-5
Conservation Alternative
Date: 07/07/95
Southwest Wetlands Consortium
An Association of: **DESIGNWORKSHOP** · Montgomery Watson · SWCA Environmental Consultants

Clark County Wetlands Park
Clark County, Nevada
This Map is for Planning Purposes Only. Clark County Assumes no Liability.

Figure 5.13 Conceptual plan, enhancement levels, and program elements of the Conservation Alternative.

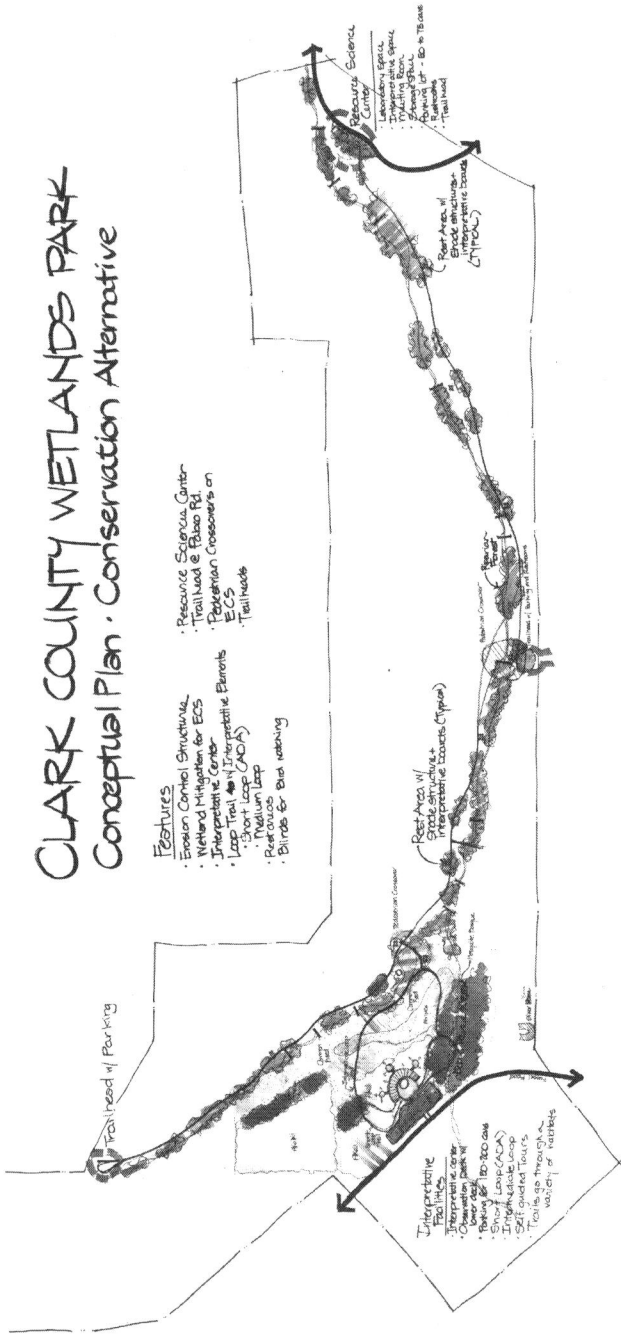

CLARK COUNTY WETLANDS PARK
Conceptual Plan · Conservation Alternative

Features
- Erosion Control Structures
- Wetland Mitigation for ECS
- Interpretive Center
- Loop Trail · w/ Interpretive Elements
  - Short Loop (ADA)
  - Medium Loop
- Restrooms
- Blinds for Bird watching

- Resource Science Center
- Trail head @ Pabco Rd.
- Pedestrian Crossovers on ECS
- Trailheads

Trailhead w/ Parking

Rest Area w/ Shade structure + Interpretive Boards (Typical)

Rest Area w/ Shade structure + Interpretive Boards (Typical)

Resource Science Center
- Laboratory Space
- Interpretive Space
- Meeting Room
- Storage/Office
- Parking Lot · 80 to 75 cars
- Restrooms
- Trailhead

Rest Area w/ Shade structure + Interpretive Boards (Typical)

Interpretive Facilities
- Interpretive center
- Observation park w/ lower deck
- Parking lot · 150-300 cars
- Short Loop (ADA)
- Intermediate Loop
- Self-guided Tours
- Trails go through a variety of habitats

Figure 5.13 (continued)

Figure 5.13 (continued)

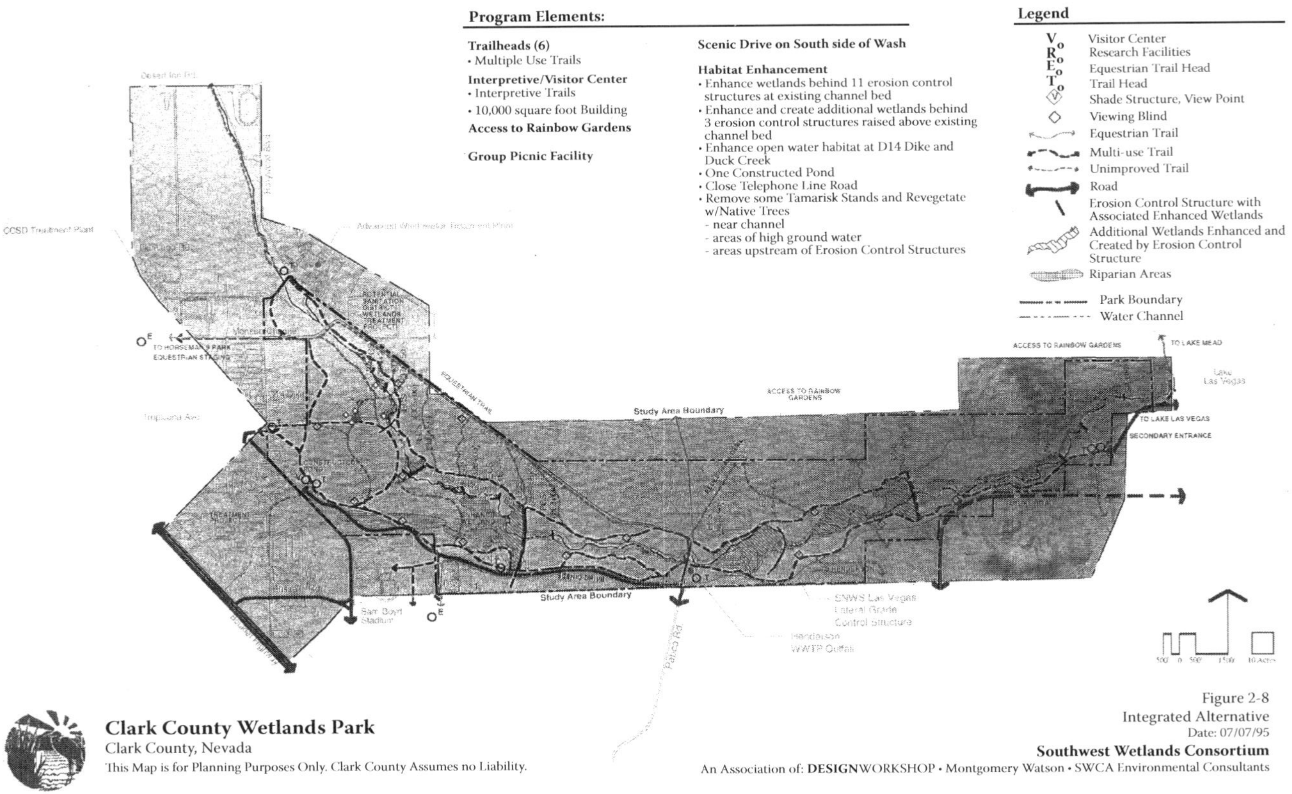

Figure 5.14 The Integrated Alternative plan for the Clark County Wetlands Park.

All the options were evaluated based on environmental and economic criteria and assessed with considerable input from agencies, government, and public users. The final master plan (Figure 5.15) addressed concerns identified in these review sessions and also resolved conflicts identified during the programmatic environmental impact statement evaluation process.

The master plan was phased to respond to the facility, budget, and scheduling needs of each erosion-control structure or park improvement over the next decade and a half. The most severe erosion occurs in the eastern (downstream) region of the Park. The master plan looked at creating longer-term fixes for this location where access is limited. As a result, only a few trails are designed with no other programmed activity. For the central region of the Park, the Pabco Road erosion-control structure was used to provide trailhead access to federal land with a middle zone of intensive public use, including a group picnic facility and location for the intended visitor center. The western region of the Park was left as a nature preserve. Because the reaches of the Wash in this latter region have not experienced severe erosion, there is more opportunity to actually enhance the vegetation and habitat without much significant work, such as having to recreate the soil necessary to establish plants.

A trail network will be built through the Park (Figures 5.16, 5.17). It is important to note that hundreds of trails currently exist as the legacy from all-terrain vehicles (ATVs), which are now banned from the site (this is also the only interest group that remained unhappy about the final plan). Concerned trail enthusiasts from the equestrian, birding, hiking, and mountain biking stakeholder groups were brought into the planning process. A few members from a naturalist group questioned the need to put a trail on both sides of the Wash, believing that it would cause too much impact, until the overall trail strategy was better explained. By taking the 200 former ATV trails and consolidating them into two major pedestrian routes with all the informal trails disappearing through time or purposely being vegetated, there will be much less impact to wildlife than the current condition. The planning team believed the desert Wash was a special place that needed to be accessible to everyone, even during the summer when temperatures approach 35°C and humidity is high. To expect the majority of people to get out of their cars and experience the Wash at this time of year was thought to be asking a lot, so vehicular access will be provided along a single road at the southern edge to enable people to enjoy the Wetlands Park without the physical need to walk through the site.

The final documents for the master-planning effort describe the planning processes utilized, the alternative implementation measures

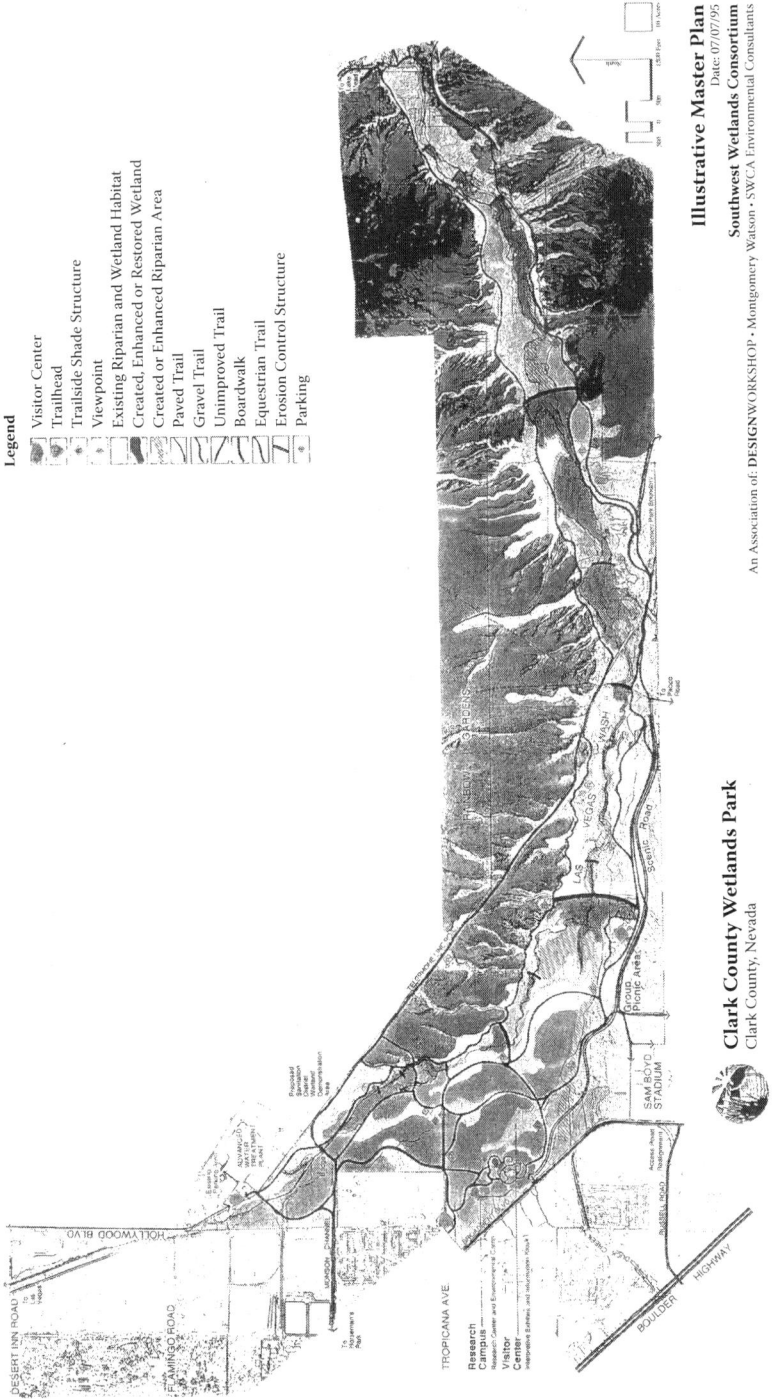

Figure 5.15 The final illustrative master plan for the Clark County Wetlands Park.

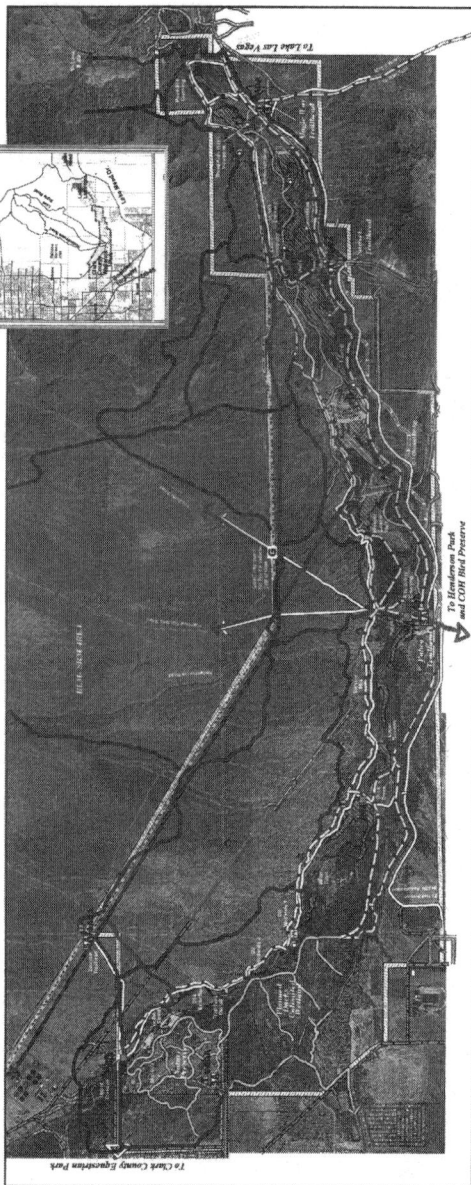

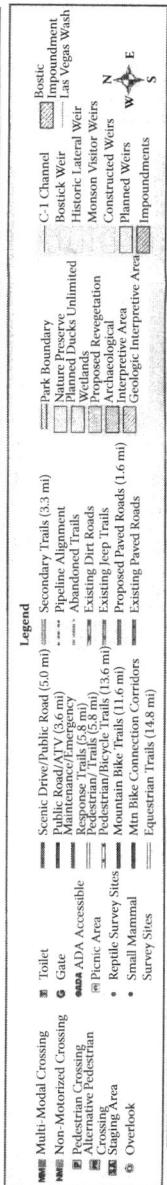

Figure 5.16 Proposed trails and facilities plan for the Wetlands Park.

Figure 5.17 Early constructed trails and trailside amenities.

considered, the justification for future consideration, description of the master plan in a sufficient level of detail, discussion of the design effort, and presentation of the environmental analysis to meet expectations of the vision. The documentation goes beyond just describing the vision; it includes information about funding, operations, and maintenance, and about phasing. Great detail, for example, is spent on outlining a strategy of land acquisition and the framework needed for permitting and environmental assessment. As such, the master plan becomes a document that is

Figure 5.17 (continued)

expected to have a long shelf-life. In 1997 it was awarded a Communications Prize from the American Association of Landscape Architects.

The implementation strategy section of the master plan is particularly instructive in its recognition and insistence that "the most important component in making the transition from a plan-on-paper to a built project is funding." The master plan goes on to identify four key elements of an effective fundraising program:

1. A phasing and carefully-developed funding plan and budget that details target items to ensure success.
2. Exploration of various approaches to establish a fundraising/ implementation organization by either designated in-house Agency staff, a partnership program of shared responsibilities, or a third party.
3. A plan itemizing the advantages, disadvantages, and demonstrated examples for obtaining actual fundraising from a myriad of sources such as local, state and federal governmental sectors, corporate, philanthropic donations, as well as a whole suite of innovative targeting approaches.
4. The often overlooked aspect of grant administration including progress reports, site inspections by donors, and other follow-up activities.

Likewise, the section of the master plan concerned with how the Park should be managed is a showcase in its presentation of a detailed listing of the interrelated duties needed to operate such a large public landscape. Operational considerations, such as routine and remedial maintenance, user safety and risk management, programming and events, stewardship and enhancement, user conflict resolution and carrying capacity, protection of abutting private property interests, and design review are elaborated upon as are a whole suite of topics related to visitor projections and consequent projected staffing requirements.

During development of the master plan, public input continually focused upon the need to develop a strong educational/interpretive program and accompanying facilities for promoting understanding about the historical development of the Wash and its significance as a conservation resource. The underlying belief throughout the master plan is that developing a comprehensive interpretive program is critical to the rehabilitation of the wetlands in the Las Vegas Wash and the recreational opportunities of the Clark County Wetlands Park. Specific elements such as interpretive trails, signs, and brochures are discussed, as well as outreach programs and special event programming. Topics identified as being appropriate for the interpretive program include public awareness about stormwater runoff, wastewater treatment, and regional water quality and quantity issues; instruction about local geology, soils, erosion, and the evolution of the Las Vegas Valley; information about the subtle ecological relationships in desert climates; a place for seminars and research activities; open space for passive recreation, education amenities, and observation of the natural ecosystem; an interpretive center with displays, storyboards, and plant and animal identifications and natural history interpretations; and a framework to educate about the human history of the Wash.

The goal of the programming was to create and enhance a land-use mix that could be enjoyed by future generations, be integrated into the desert surroundings, provide respite from the city, and balance public use and habitat enhancement. Existing conditions relating to views, vegetation, and hydrologic conditions all created the framework for the site design. And the major spirit guiding site design was inspired by the movement of water. Each introduced element (Figure 5.18) was selected to contribute to the visual continuity of the entire Wetlands Park. For example, because of climatic conditions, particular attention had to be paid to the use of materials, for example, the palette colors representing desert hues and use of metal roofs to help minimize maintenance. It was important to establish a sense of place, so the overall design goal was to create a landscape that has a timeless spirit that would reflect the characteristics of this unique landscape; in other words, a wetland in the Mojave Desert at the edge of the fastest-growing city in the United States.

Each gateway provides an opportunity to identify the Wetlands Park and to assist in orientation for park facilities (Figure 5.19). Budgets were

Viewing Areas

Viewing Blind

Figure 5.18 Sketches of representative viewing areas and blinds.

put together for each of the proposed trailheads (Figure 5.20). Trail nodes were designed to provide visitor rest stops, interpretive education opportunities, and scenic vistas or windows into or out of the Park. New trails will be positioned to bring visitors to specific areas or to link up to other recreational opportunities on adjacent lands and to provide an educational storyline to all the habitat types within the wetland-wash. A clear hierarchy of trails (Figure 5.21) will reduce the impacts of the numerous pedestrian paths and vehicular trails that currently crisscross the site. A set of proposed trail guidelines was produced that considered the following elements: level of access, trail widths and rights of way, maximum percentage of slope, shoulder-clearing height and width, alternate surface construction material (Figure 5.22), and trail uses for each particular type of trail. These recommendations were used to produce a series of schematic detail drawings and a rough order of magnitude construction budget of about $13 million.

The Clark County Wetlands Park is planned to function as a classroom with organized and informal education programs located throughout the site. A system of interpretive and graphic signage will provide a sense of the cultural heritage and appreciation of the natural environment

Conceptual Sketch of Sunset Park Trailhead

Figure 5.19 Conceptual sketches of trailhead facilities.

and understanding of the recent history of the Wash and the impact that humans have had on the landscape.

Although the public is always in need of sites for active recreation such as ball fields, the emphasis was placed on developing recreational opportunities that are compatible with the conservation and restoration of the Las Vegas Wash. Waterfowl viewing blinds will be strategically placed for the birders and picnic areas sited at some of the gateway locations. Locations for hiking, horseback riding, birding, photography, and mountain biking will be established. Because the Wash carries treated wastewater, trails will be sited to discourage interaction with the water. Attention was directed to making visual connections to the water as an aesthetic amenity while making clear that other recreational activities, such as swimming, fishing, or kayaking will not permitted.

Clark County Wetlands Park
Potential Pabco Trailhead Costs

Example Civil Engineering and Design Budget-1 Acre Trailhead

ITEM	SIZE	QUANTITY	UNIT	$/UNIT	TOTAL
Landscape Design				Lump	$5,000
Civil Engineering				Lump	$10,000
Earthwork				Lump	$5,000
Solar Power Set Up				Lump	$10,000
Utilities (emergency phone)				Lump	$5,000
Water Hook Up		1	each	$10,000	$10,000
Total Proposed Costs					**$45,000**

Hardscape Materials

Composting Toilets		1	each	Lump	$50,000
Maintenance Building		150	sq.ft.	$50	$7,500
Access Gates		3	each	$1,000	$3,000
10' × 10' Ramadas		2	each	$1,500	$3,000
Seating Walls	18" Wall	100	lin.ft.	$100	$10,000
Upgrade Pabco Crossing		1	Lump	$300,000	$300,000
Parking Lot	150' × 150'		Lump	$20,000	$20,000
Benches		5	each	$400	$2,000
Picnic Tables		5	each	$400	$2,000
Information Kiosk		1	each	$5,000	$5,000
Trail Head Signs		4	each	$150	$600
Wetland Boundary/Warning Signs		30	each	$50	$1,500
Interpretive Trail Signs		5	each	$2,000	$10,000
Park Entry Sign		1	each	$6,000	$6,000
Trash Containers		5	each	$100	$500
Group Shade Ramada		1	each	$10,000	$10,000
Wildlife Viewing Platform		1	each	$30,000	$30,000
Total Proposed Costs					**$461,100**

Landscaping

Landscaping		1	acre	$15,000	$15,000
Turf and Wildflowers		1	acre	$15,000	$15,000
Irrigation System		1	acre	$2,500	$2,500
Total Proposed Costs					**$32,500**
Total Proposed Pabco Costs					**$538,600**

* All Costs are Example Figures Only and Should Not Be Used for Construction

Figure 5.20 Detailed comprehensive budget for construction of a trailhead.

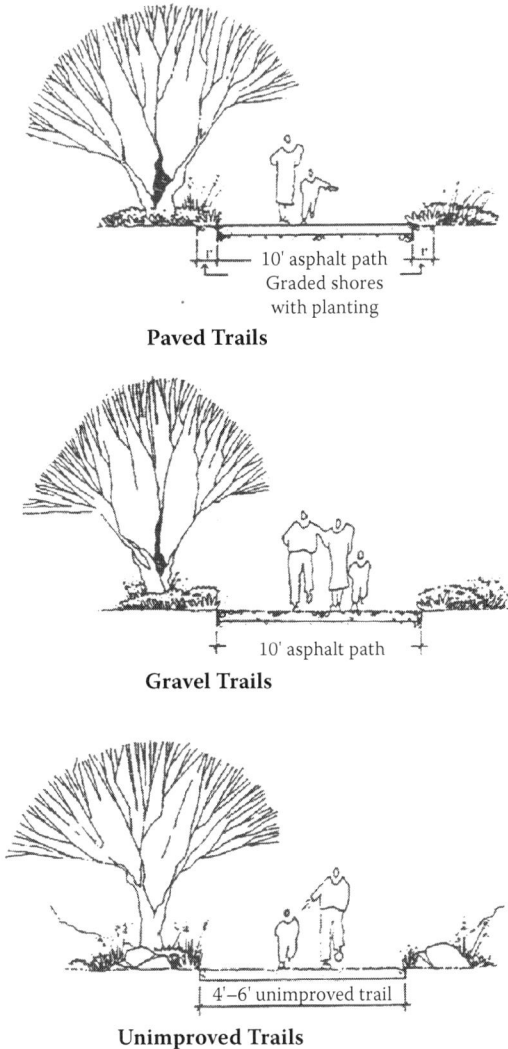

Paved Trails

10' asphalt path
Graded shores
with planting

Gravel Trails

10' asphalt path

Unimproved Trails

4'–6' unimproved trail

Figure 5.21 Schematics of trail typology and hierarchy.

From the start, a rich diversity of plant life was recognized as critical for the future of the Wetlands Park in terms of creating a higher quality of wildlife habitat. One of the most important elements of the master plan therefore provided a roadmap for the protection, renewal, refinement, and embellishment of the Las Vegas Wash by enhancing the habitat for the many species that call the area home. In order to accomplish this objective, special attention had to be focused on the complex dynamics of water on the site.

Mountain Bike Trail

Secondary Trail for Hiking

Multi-Use Equestrian Trail

Pedestrian/Bicycle Trail

Figure 5.21 (continued)

Clark County Wetlands Park
Trail Surface Unit Cost

Surface Material	Cost Per Mile/Longevity	Advantages	Disadvantages	Suggestions
Concrete	$300,000–$500,000/long term (20 plus years)	Lowest maintenance, hardest surface, supports multiple use, resists freeze/thaw, most resistant to flooding	High initial cost, unnatural, expensive, subject to cracking, need vehicle access to construct and repair	Allow vegetation to grow over edges for natural appearance, crown slightly for drainage
Asphalt	$200,000–$300,000/medium long (7–15 years)	Hard surface, supports most types of use, all weather, smooth surface to comply with ADA guidelines, low maintenance	High initial cost, unnatural, expensive, subject to cracking, tacky in hot sun	Allow vegetation to grow over edges for natural appearance, crown slightly for drainage
Granular Stone	$80,000–$100,000/medium long (7–10 years)	Soft but firm surface, natural material, moderate cost, accommodates multiple use	Surface can rut or erode with heavy rainfall, regular maintenance needed to keep consistent surface, not for areas prone to flooding or steep slopes	Can utilize imploded building waste as base and surface
Boardwalk	$1.5–$2 million/medium long	Necessary in wet or environmentally sensitive areas, natural looking surface, low maintenance, supports multiple use	High installation cost, costly to repair, slippery when wet	Good for wetlands, can also use floating dock type material as boardwalk

	Cost			
Wood Chips	$65,000–$85,000/Short term (1–3 years)	Soft but firm surface, natural material, moderate cost, accommodates multiple use.	Can become soggy in poorly drained areas, requires replenishment.	Hardwood chips are most desirable. Avoid material with sharp and angular chunks due to dull chipping machines.
Native Soil	$50,000–$70,000/Short to long depending on local use and conditions	Natural material, lowest cost, low maintenance, can be altered for future improvements, easiest for volunteers to build and maintain.	Dusty, ruts when wet, not an all weather surface, can be uneven and bumpy, limited use, possibly not accessible.	Good for wilderness trails and areas where good compact soils are present.
Resin-Stabilized	Cost varies depending on type of application	Aesthetics, and less environmental impact, possible cost savings if soil used, can be applied by volunteers.	Need to determine site suitability and durability, may be more costly in some cases.	Good for wilderness trails and areas where good compact soils are present
Recycled Material	Cost and life vary	Good use of recycled materials, surface can vary.	Design appropriateness and availability may vary.	

* All unit costs were derived from "Trails for the Twenty First Century," Rails to Trails Conservancy
* These are good base unit costs; price per mile will vary depending on each particular site, trail width, and volunteer efforts.
* These cost do not include trail elements such as landscaping, lighting, culverts and bridges.

Figure 5.22 Unit costs of various trail surfaces.

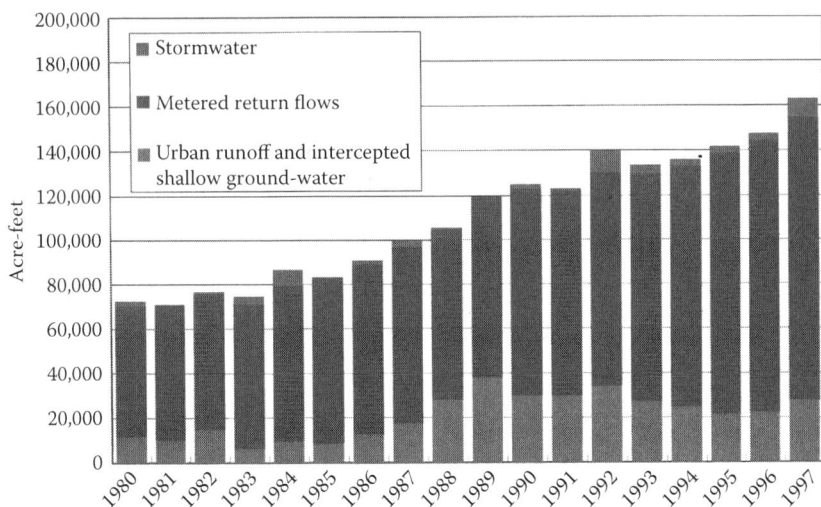

Figure 5.23 Breakdown of annual flows in the Las Vegas Wash from 1990 to 1997.

Environmental problems and ecological rehabilitation

The rehabilitation and utilization of the Las Vegas Wash has become one of the most important environmental issues in southern Nevada. Water is very much both the architect and the vandal in this story. It is because of the water in this extremely arid environment that the place is so special; with only four inches of rain a year, the site is really one of the driest places in the United States. Before the urbanization of the Las Vegas Valley, the Wash was an ephemeral stream, only flowing during periods of significant rainfall. Today, four sources of water feed and channel through the Las Vegas Wash: metered flows of wastewater effluent, groundwater, general urban runoff, and stormwater-produced runoff (Figure 5.23). In a way, the site is the physical manifestation of the old adage that "it all comes out in the wash." The predominant water source—90 percent of the water during dry weather—is wastewater effluent (Figure 5.24). The other 10 percent is provided by groundwater rising to the surface and urban runoff from Las Vegas. This water is slightly saline and contributes a salt load to Lake Mead and the Colorado River. During the few but very intense desert storm events, pulses of stormwater that enter the system absolutely dominate and wreak serious damage. Therefore, whereas the dry weather flows (from urban runoff and wastewater) destabilize the existing channel because it is not in equilibrium with its slope, it is the stormwater pulses that exact the catastrophic damage.

Figure 5.24 Treated wastewater flows on the way to the Wetlands Park.

Water quality becomes very much a management issue and a health issue for all (Figure 5.25), including the flora and fauna. Wastewater in the Wash is tertiary treated effluent and released by permit through the State of Nevada. The effluent is filtered and disinfected, and much of the contaminant load, including nutrients, biological oxygen demand, total suspended solids, and coliform bacteria, is removed. Still, the cocktail of treated wastewater and runoff has raised concerns about the quality of the city's drinking water. While this has been an issue for decades, concerns were elevated by an accident elsewhere in the United States in 1994 that left over thirty people dead from cryptosporidium bacteria, and the recent discovery of fish with deformed and dysfunctional endocrinal systems where the Wash empties into Lake Mead.

The nonpoint source urban runoff is another situation altogether. Because of people overwatering their lawns, the Wash water is high in herbicides, fertilizer nutrients, total dissolved solids, and salts. The

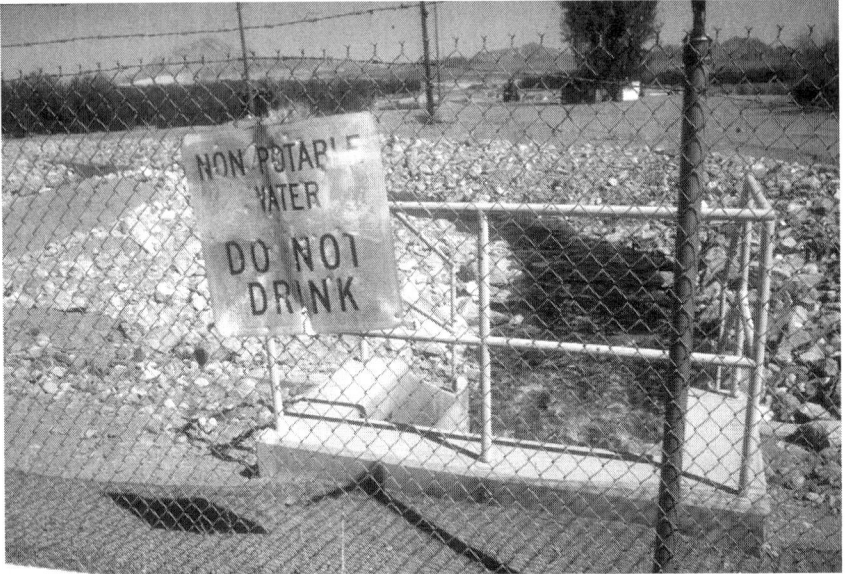

Figure 5.25 Wastewater signage reminding people about nonpotable source of water in the Wash.

groundwater is also high in salt concentration and, particularly in the middle of the Wash, elevated perchlorate levels. Stormwater, however, is the biggest challenge as another nonpoint source that has a high oxygen demand, as well as elevated levels of suspended sediments, salinity, nutrients, and bacteria. With so few storm events in any given year, between them there is an accumulation of all the pollutants—the sins of the urban environment—so that during the first flush of a heavy rain, a polluted mixture courses off the streets and enters the Wash. Indeed, were it not for the constant discharge of treated wastewater to dilute the urban runoff, the overall quality of the water in the Wash would be much worse.

The good news is that the Las Vegas Valley water-quality agencies have been taking steps for several years to clean up the point and nonpoint pollution sources of stormwater runoff that the Wash receives by employing such best management practices as maintaining catch basins and storm drains, street sweeping, eliminating illegal connections, removing hazardous waste, controlling litter, enforcing local ordinances, monitoring stormwater, and utilizing public education, such as having school children stenciling little fish near all drains, identifying that they connect to Lake Mead, the source of drinking water for the area.

The rapid population growth of Las Vegas has resulted in severe damage to the Wash. In 1950, there were about 50,000 people living in the

Valley, but once air-conditioning became affordable, exponential growth occurred. Those 50,000 people produced about 5 million gallons per day of wastewater flow. Today, half a century later, the almost 1.5 million inhabitants produce over 160 million gallons per day of wastewater which goes down the Wash. However, this is not nearly as much of a problem from an erosion perspective as is the runoff from stormwater events. The increase in impervious surfaces coincident with the city's expansion has exacerbated the runoff and erosion problem (Figure 5.26). Naturally severe storm events in a desert environment, which once would have infiltrated the ground, now rush off the roofs and streets causing serious erosion in the receiving Wash. What was once a shallow and broad meandering stream channel has become a narrow and steep-sided trench (Figure 5.27). An average of over 100,000 tons of sediment is lost to Lake Mead annually, an incredible rate of about 50 to 1,600 tons per day. Las Vegas Bay, the inflow bay in Lake Mead, is becoming eutrophied and losing its oxygen because of the incursion of nutrients and sediments. As a result, the U.S. National Park Service has become extremely worried about the upstream urbanization in the Valley. The sediment contains trace contaminants that now find their way into urban water supply systems in southwest Arizona and California, not to mention Las Vegas itself.

Las Vegas is one of the fastest growing metropolitan areas in the United States, its population growing by an incredible 85 percent during the 1990s. This breakneck growth has had serious consequences on water management concerns in the region, leading some to refer to the city as "the water-guzzling Sodom on the Colorado [River]." The Colorado River is southern Nevada's primary source of water. Las Vegas's culture of "water waste" is actually encouraged by an arrangement whereby the more water that flows through its wastewater treatment plant, progresses through the Wash, and enters Lake Mead, the more water that can be given back to the city as "return flow credits." In other words, the city is allowed to reclaim and proportionally draw the equivalent amount of water from Lake Mead that is sent into it as treated wastewater. Given that the degradation of the wetlands in the Wash is a result of excess water input, the situation of reclaiming the damaged landscape looks bleak from a strictly policy point of view; clearly, some structural improvements are required.

In this respect, a landscape architecture student who studied the site believed that though much is made about the open-space potential offered by the Clark County Wetlands Park, it is important to realize that its most significant value is for water management because the Wash is actually an engineered system and the latest development in a long history of hydrological manipulation in the Las Vegas Valley. It becomes impossible with this perspective to interpret the Wetlands Park as anything other than a political landscape.

Figure 5.26 Typical drainage channels in Las Vegas that carry stormwater pulses directly into the Wash.

Figure 5.26 (continued)

Stormwater runoff, however, does not count as return-flow credits and, as such, is an unutilized potential resource. A network of detention basins has been constructed or are planned throughout the Valley (Figure 5.28) in order to capture the water temporarily and slow down its rate of movement into the Wash, thereby contributing toward reducing flooding and erosion. Because of the design of the existing basins, however, they cannot be used to store stormwater for conservation reuse. As a result, alternative options are being examined to convert some of the existing basins and to make sure the new basins are able to retain and treat, not merely detain, stormwater before its release into the Wash (because of local variances, it is currently impossible to legally use untreated stormwater for aquifer recharge).

If ever there was a signature case study that clearly demonstrates the need not only to identify problems early but to control them, it is here. In 1981, for example, an early master plan stated that erosion-control measures should be implemented immediately to prevent head-cutting (the upstream migration of massive channel entrenchment) in order to protect the upper Wash and to stabilize the lower Wash. At that time, this was estimated to cost about $200,000 for the installation of four erosion-control structures. Today, about twenty-two structures are needed at an estimated cost of $60 million. The message here is clear: Take care of problems as they are discovered.

Figure 5.27 Views of the Las Vegas Wash from the Northshore Road in (from top to bottom) 1975, 1985, and 1999.

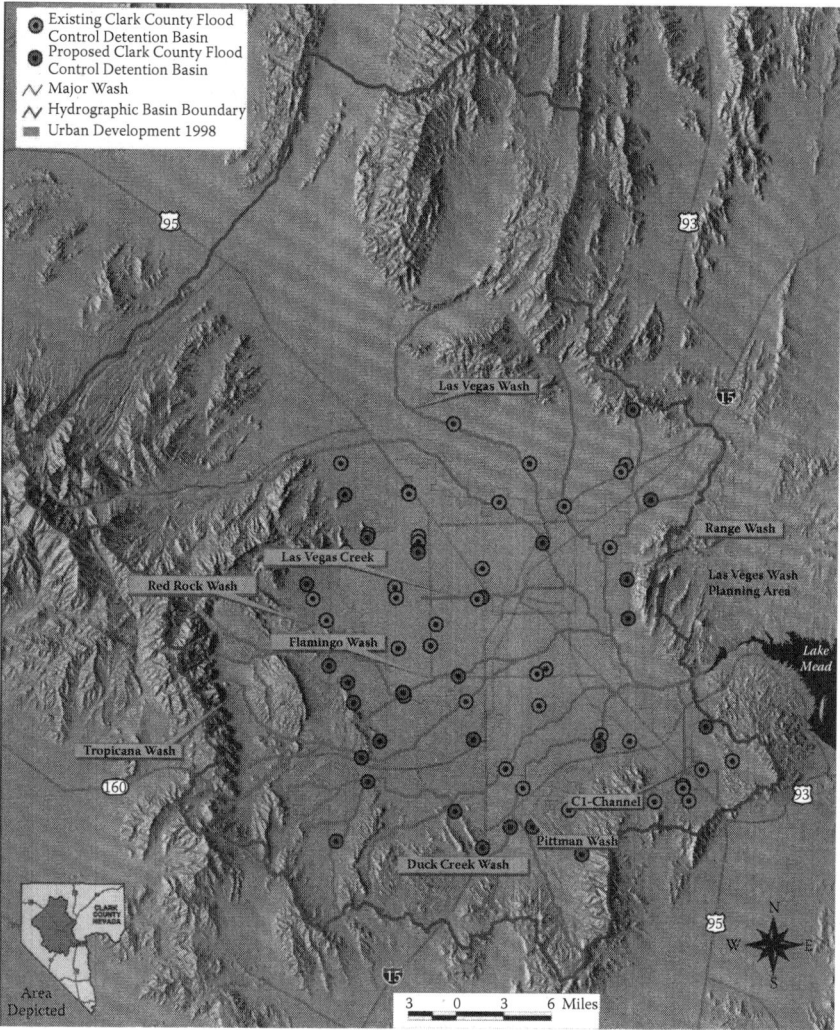

Figure 5.28 Location of existing and proposed stormwater detention basins in the Las Vegas Valley.

The current call to action for erosion control necessitated extensive and new hydraulic modeling based on a much better understanding of the fluvial and geomorphological processes of the Wash. These background data were required to define the erosion-control needs, which led to the ability to design and construct the stabilization measures, which in turn led to an adaptive assessment of sediment movement dynamics.

Local agencies predict that wastewater flows in the Wash will increase by approximately 60 percent over the next few decades. Futures modeling used a base flow (600 cubic feet per second) of about double today's base flow in order to predict the implications of the ultimate projected build-out in the Las Vegas Valley over the next two decades. Then by examining the history of previous 100-year-storm patterns, it was predicted that if nothing were done to attend to the erosion problem, 2.3 million tons of soil would be lost over the next 20 years, resulting in an additional 10 to 35 feet of headcutting and leading to catastrophic effects.

Efforts are now actively underway promoting channel stabilization and limiting channel downcutting and reducing bank erosion, armoring the channel with vegetation, balancing sediment transport and deposition, and restoring the lost ecosystem. A host of factors need to be considered throughout the engineering process of stabilizing the Wash: wastewater and urban flow, flood discharge and seepage, topography, soils and geology, slope and alignment, vegetation, water quality, wildlife protection and enhancement, predicted future conditions, ownership and access, permitting, recreation needs, implementation and maintenance costs, funding, constructability, and risk, and public involvement.

To stabilize the Wash channel bed from further downcutting, the Water Quality Advisory Committee considered a series of concepts. The present channel bed ranges in slope from 0.22 to 0.80%, considerably steeper than the slope of 0.15 to 0.35% necessary for stability. This translates into the requirement of about 157 vertical feet of stabilization needed to achieve the equilibrium slope. Given this requirement, the entire Wash was examined to determine where best to locate the channel gradient control weirs. Where conditions were found to be favorable in terms of maximizing sediment storage and wetland creation, the installation of erosion-control structures was planned, the design of each such structure tailored to the site specifics and budgeting estimates. The particular selection of erosion-control structures (Figures 5.29, 5.30) also considered technical issues such as discharge capacity, force resistance, bedload scouring, foundation flexibility, seepage and piping protection, overturning and sliding stability, and earthquake safety, in addition to design issues such as aesthetics.

Figure 5.31 shows a site analysis of the erosion sensitivity of the Wash and the locations identified for establishing the twenty-two erosion-control structures deemed necessary to keep erosion in check. Because of the presence of highly erosive soil materials, the Wash is now two to four times as steep as it should be in order to preserve its stability. A major effort is therefore needed to restore a gradient of no more than about 16 feet per mile along the complete length of the Wash. Pilot bank stabilization projects are underway, the performance monitoring of which will lead to design improvements for the ongoing erosion control program. The wildly ranging flow depths and discharges complicate the selection of effective erosion-control structures. Certainly, this is very much a situation where

Figure form Kennedy/Jenks/Chilton, 1989

Figure from Kennedy/Jenks/Chilton, 1989

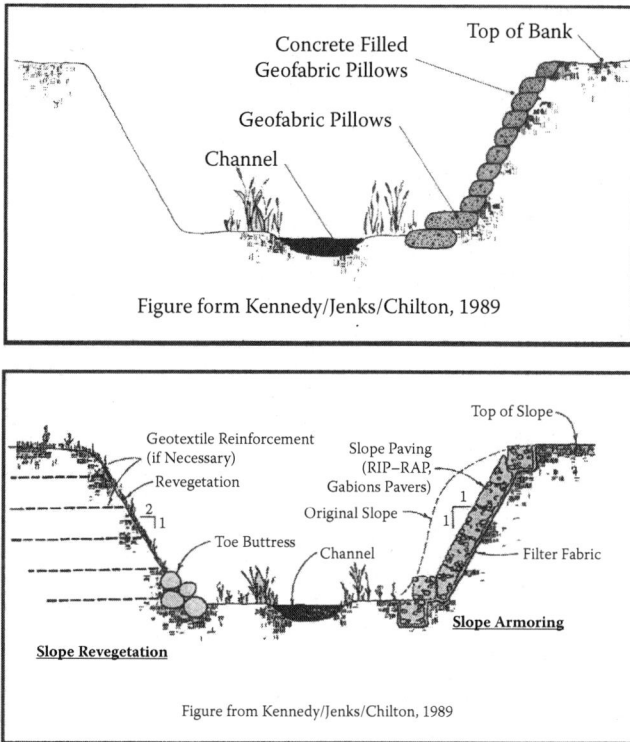

Figure 5.29 Conceptual drawings of prototype erosion-control structures for the Las Vegas Wash.

a single prototype design would be inappropriate. Instead, a mixture of hard engineering and soft bioengineering techniques were adopted, the strategy being to stabilize channel headcutting and hold the channel bed in place, thereby allowing for a vertical drop in channel grade to be accomplished without the headcutting advancing upstream. Both simple interim structures of riprap (large rock) and high-capacity concrete structures are being used. Earlier stabilization structures used rocks of insufficient size in relation to the extremely high-energy environment of the Wash, the result being that many were washed away after the first significant storm event. During 2001, over 2 miles of recycled, large-sized concrete rubble was graded and placed in the Wash to stabilize the bank toe, which remains successfully in place.

The first erosion-control structure was built at Pabco Road. It is an 816-foot long and 7.5-foot high rock gabion with a concrete cap. Although this makes it more expensive than what a normal, uncapped structure would have cost, as this was the first one to be built, it had to sustain the

Figure 5.30 Early erosion control installation projects.

full brunt of the erosive forces in the Wash before other supporting structures could be erected. Visually, the white concrete capping in a tan desert
setting is not a perfect aesthetic, so it was not the best visual choice of
materials. The foundation is composed of a geotextile fabric lining, over
top of which is a gabion and steel sheet pile seepage cutoff system, with a
total cost of $4.6 million.

The next erosion-control structure that was constructed, referred to
as the Historic Lateral, was not as substantive. It is a confined rock riprap
structure, 609-feet long and 6-feet high that was designed to be more natural looking. Built over top of the same foundation as the Pabco Road structure, it cost $1.7 million and has since filled in with willow and cattails.

The third erosion-control structure put in place was an interim structure referred to as the Demonstration Weir. It was built in a crook of a
bend of the Wash, in a narrow area of exposed bedrock suitable for holding rubble, and with an area of sediment in front of the bend that provided a good location for wetland development. The weir was 18-feet high
and 400-feet long, formed of 24,400 tons of unconsolidated rubble, with no
foundation, and at a cost of $250,000. Designed as the toe for another per-
manent structure to be built later immediately upstream, this weir, which
was designed to resemble a natural riffle, is a good example of sustainable
design in that the large material placed instream was recycled. Las Vegas
is famous for imploding massive buildings. So when the El Rancho Hotel

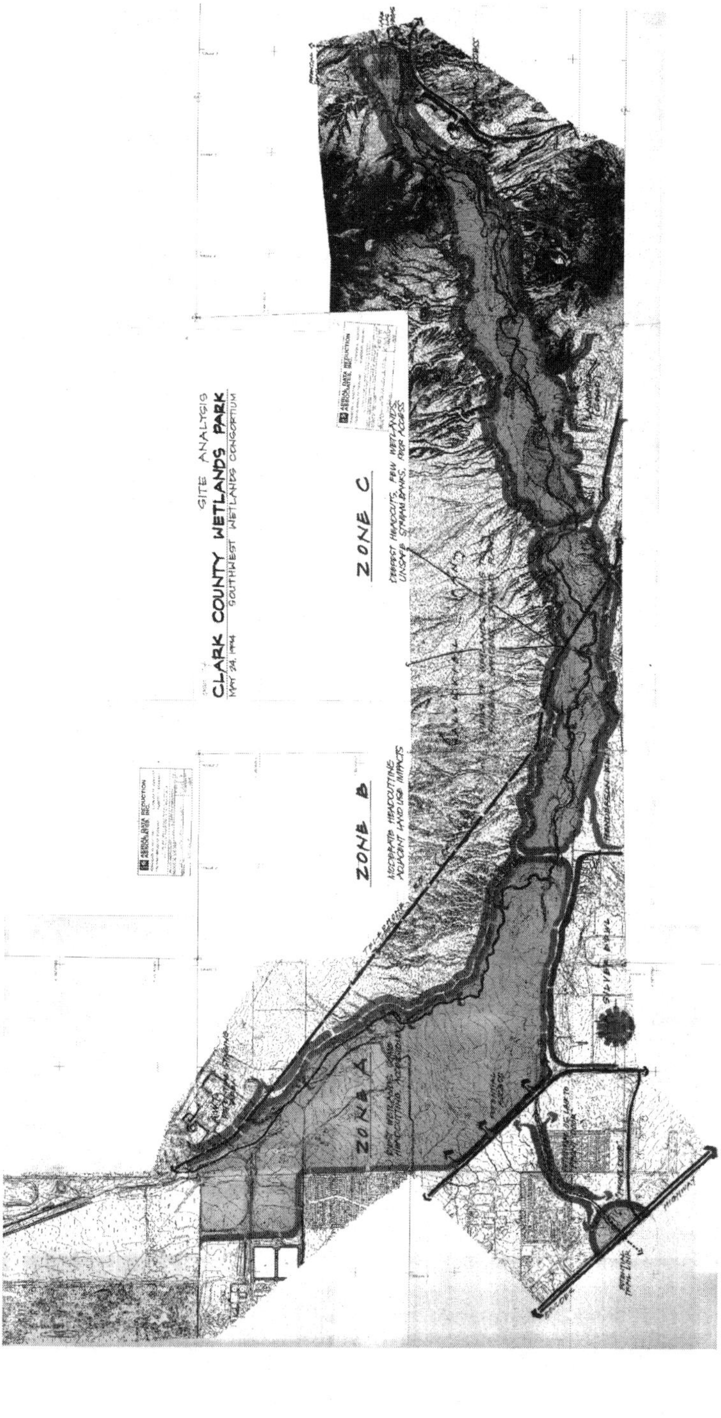

Figure 5.31 Location for construction of erosion-control structures.

and Casino was imploded on the Strip, an enormous amount of concrete was stockpiled and saved for use in the erosion-control structures. This became one of the more innovative moments of the restoration process: taking a piece of the actual Las Vegas Strip and moving it down to the natural environment of the Las Vegas Wash and observing its effectiveness in checking erosion. Over 2 acres of wetland habitat were created behind the weir.

The Fire Station Weir was a more modest affair. It was formed by placing boulders that helped create a 7-acre wetlands mitigation project and decreased the current flow before a pronounced S-bend located downstream. In another demonstration of sustainable design, the weir takes the sediment and gravel out of solution and this material is actually mined and recycled for use in other projects located in the lower Valley.

The goal of the erosion-control measures was to "train" the river such that flows were directed to specifically armored and thus protected locations. Specifically, one aspect of the task was to sculpt or lay back the slopes and toe in order to deflect water. Spur dikes were used for mechanical protection to shunt the river around the corner. It is important to reiterate that the Wash is an area with a rich cultural heritage, and that in fact over 80 percent of the erosion-control sites will be constructed in locations of archeological significance (in one case, an 8,000-year-old campsite) that will need to be catalogued and removed.

The Southern Nevada Water Resources Authority, summarizing 15 years of experience in installing eight erosion-control structures of variable levels of complexity in the Wash, generated a set of observations and lessons learned, some of which follow:

- Lack of seepage foundation rapidly leads to foundation failure.
- Downstream scour protection is necessary to prevent structure toe failure.
- Small diameter (< 18-inch) rock structures cannot withstand flood forces.
- Incorporation of bioengineering approaches increases the resistance of the structures.
- Use of recycled concrete rubble in place of rock riprap is highly recommended.
- Costs for installing and maintaining rock or recycled concrete structures compares favorably to more substantial concrete and steel weir designs.

The last component of the rehabilitation process is to go back and establish about 160 acres of wetlands in the back-flooded areas (Figure 5.32). The Las Vegas Wash Wetland Enhancement Project is designed to actively introduce desired native wetland and riparian

Figure 5.32 Plans for restoring channel wetlands and riparian habitat.

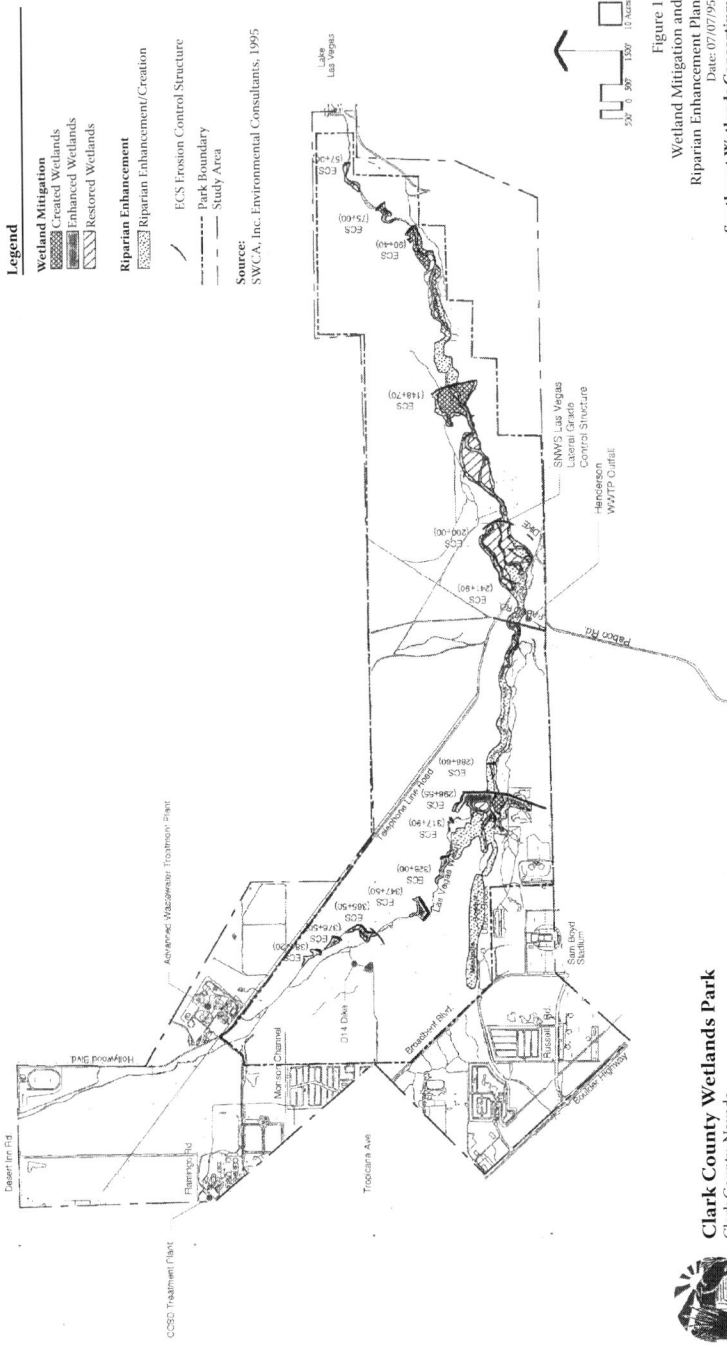

Legend

Wetland Mitigation
Created Wetlands
Enhanced Wetlands
Restored Wetlands

Riparian Enhancement
Riparian Enhancement/Creation

/   ECS Erosion Control Structure
– – –   Park Boundary
——   Study Area

Source:
SWCA, Inc. Environmental Consultants, 1995

Figure 1

**Wetland Mitigation and
Riparian Enhancement Plan**
Date: 07/07/95

Southwest Wetlands Consortium
An Association of: **DESIGNWORKSHO** • Montgomery Watson • SWCA Environmental Consultants

Clark County Wetlands Park
Clark County, Nevada
This Map is for Planning Purposes Only. Clark County Assumes no Liability.

Figure 5.32 (continued)

Figure 5.33 Early design drawings for creation of impounded wetlands behind erosion-control structures.

plants that will attract and sustain wildlife, remove nutrients, and precipitate dissolved solids (Figure 5.33). Also, the roots will function as anchors and thereby reduce further streambank erosion. Bioengineering techniques, such as planting cottonwoods, cattails, and various willow and bulrush species into jute or coir logs were used to establish the wetland and provide immediate soft erosion control. By armoring the bed and banks of the Wash and removing the sources of sediment discharge, the overall site aesthetics will be improved for the Wetlands Park. The approach undertaken was to plan and build pilot studies, letting nature work first.

Four more structures have been designed, one of which was named after Vern Bostick, the visionary who dreamed and worked toward a

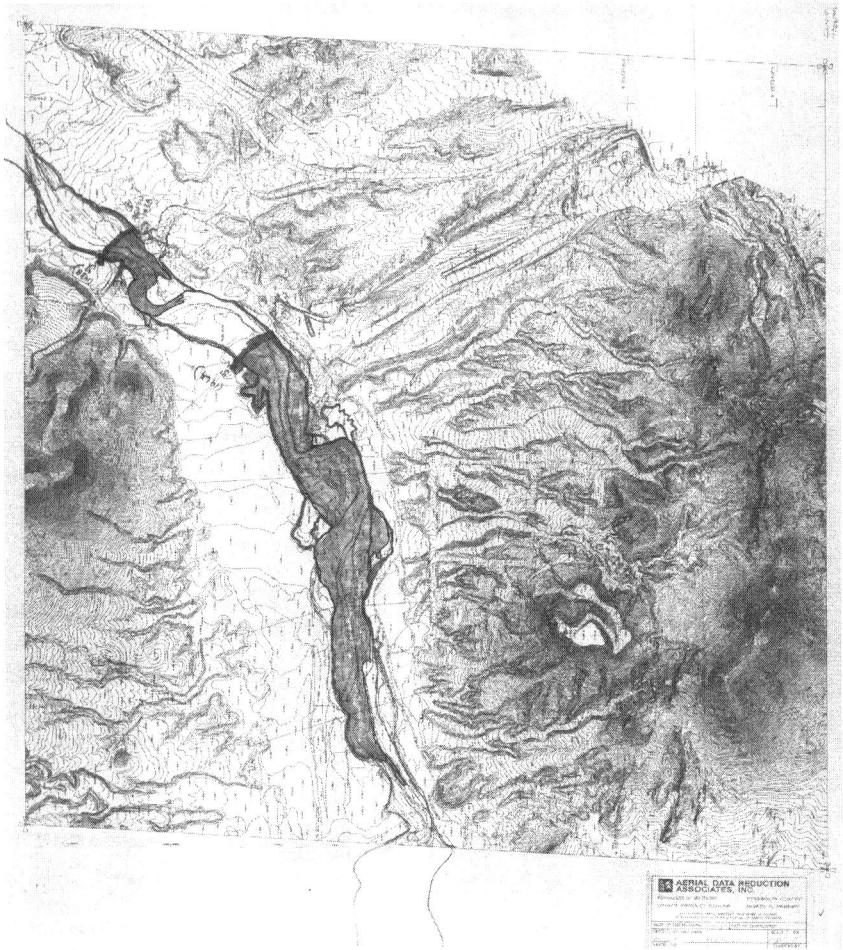

Figure 5.33 (continued)

restored wash-wetland systems for many decades. Given that more than a dozen erosion-control structures have been or will need to be constructed, a big element in the overall design process is adaptive assessment. In this respect, the site is looked upon as a living laboratory where previously built structures can be critically evaluated as to whether they are doing their job in reducing erosion. Water quality is monitored behind the erosion-control structures to see if the wetlands are performing as expected. In particular, by impounding the Wash and creating wetlands upstream of the erosion-control structures (Figure 5.34), careful attention has had to be paid to avoid situations in which oxygen levels are lowered and conditions established conducive to the development of avian botulism.

Figure 5.34 Successful wetland establishment behind erosion-control structure and use of irrigation system.

Ecologically, the premier goal is to try to outrace the onslaught of habitat degradation. With every storm event, wetlands are being lost at an alarming rate. The actual process of wetland destruction here, though, is the opposite of that occurring for wetlands located elsewhere. Worldwide, wetlands suffer most as a result of being filled in, but for the Las Vegas Wash the wetlands are actually losing the alluvial deposits that created them. As such, restoration is needed in terms of reconstructing the physical structure of the wetlands as a prelude to fostering biodiversity.

Once the channel stabilization strategy was in place, it set the stage for the other activities. The restoration actions were therefore augmented by other actions to establish the Wetland Park's value:

1. Control of the 14-mile perimeter of the Park had to be established.
2. A strong advocacy group for the trail system to help the Park enforce the regulations was created.
3. Off-road vehicle use was eliminated.
4. Native plants were reintroduced into the Wash ecosystem.
5. Data collection continued to inform the process of design, construction, and maintenance.

Construction is planned to take place over a 20-year period. As thirty agencies and organizations are part of coordinating committee, a clear road map is needed as well as knowing what is being monitored that will ultimately be beneficial toward progressing along that road map. The decision was made to concentrate on developing the visitor and interpretive areas within certain parts of the Park, leaving large portions undeveloped with minimal paths and access. Most of the Park will therefore function primarily for sustaining wildlife.

Another issues is that the wastewater that has been discharged into the Wash since the 1950s has developed a vegetation community that is not indigenous to the area. Prior to regular discharge, the xero-riparian community was dominated by dense stands of honey mesquite. During the 1960s and wetland development, cattails and the common reed became dominant. Today the brackish soils of the Wash, once inundated by the floodwaters but now left high and dry, are predominantly populated by quailbush and tamarisk, the latter being a noxious exotic plant that is the scourge of riparian areas in the southwestern United States; both are invasives able to survive in areas with a low water table. Indeed, tamarisk has increased from being about 20 percent of vegetation in the Wash in 1975 to representing approximately 80 percent of the total vegetation today. Nonetheless, because it was believed that if attempts were made to eliminate all the tamarisk in one fell swoop the major effort would also ruin the remaining useful habitat, it was decided to tackle the problem in phases.

Because there is a difference in the degree of inundation required by the undesirable exotics and the native plants in the Wash, attempts were made to capitalize on this by fine-tuning the hydrology to encourage development of the latter. Careful attention was therefore paid to optimizing the hydrologic regime and geomorphology of the reintroduction sites behind the erosion-control structures (Figure 5.35). The intent was to create a series of these native cells where there would be intensive management by removing exotics and reeds and fostering the

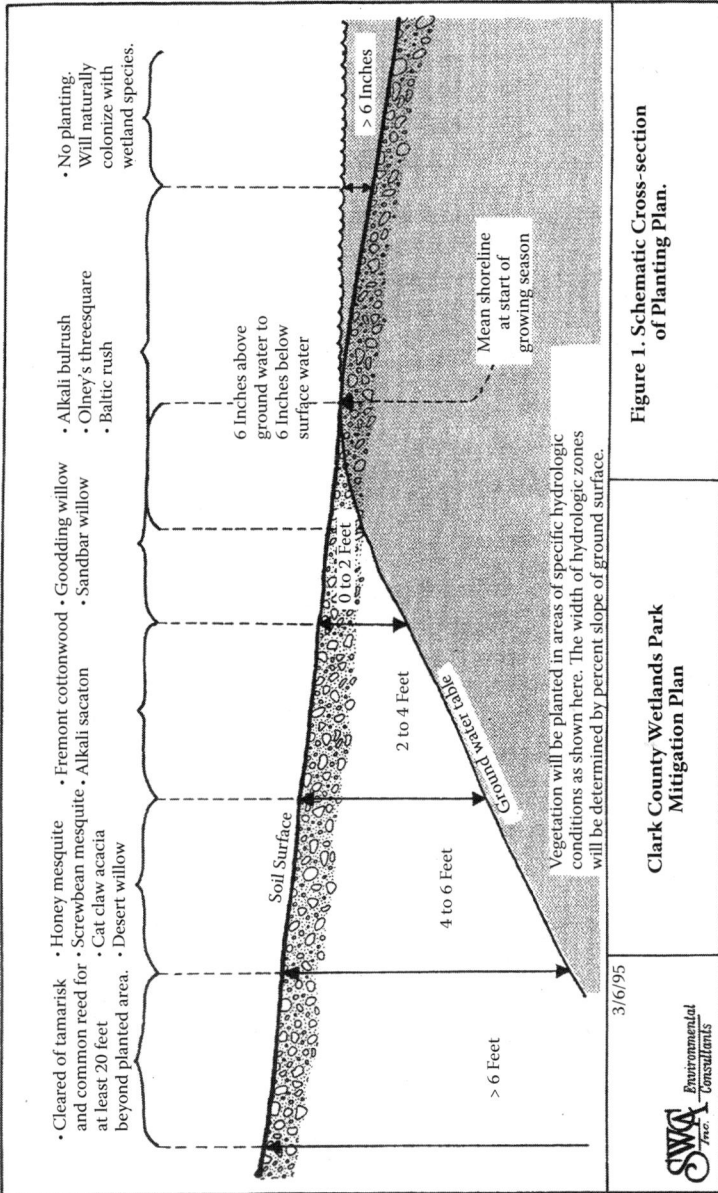

Figure 5.35 Schematic cross section of planting plan for impounded wetland.

willows and other native wetland species (Figure 5.36). The development of these sites would continue through intensive maintenance. As a result, these native cells are not really "natural systems" at this point. However, if these native plants are given enough of a head-start they can outrace the exotics and create a healthy foothold that may spread elsewhere within the Park (Figure 5.37). Furthermore, establishment of such native cells will enhance wildlife richness. The key then to the successful reintroduction of endangered animals is to successfully reintroduce native riparian and wetland habitat. To maximize the restoration efforts after completion of site development, topsoil from the access road construction site will be salvaged for use on disturbed sites to ensure a foundation for revegetation activities, and native plants from the road construction site will be saved for replanting elsewhere in the Park.

Why is there a concern about any of this in the first place? Why attempt to "restore" a wash that is largely filled with treated sewage effluent in the middle of the Las Vegas Valley? In point of fact, what can be learned from this project will likely have significance for restoration efforts elsewhere? Shrinking native riparian habitat across the south-western United States has forced many species toward an endangered status. In the Mojave Desert, although the riparian areas comprise only about 5 percent of the total land area, they support the vast majority of all species and thus play a critical role for sustaining regional biodiversity. Many of the riparian areas in the Southwest have been adversely affected by water management, which has shifted the flora toward exotic vegetation with the natural structure and floral diversity often disappearing. Surveys have shown that there is a positive relationship between the extent of native riparian flora and the density of native breeding birds. Here then, the opportunity exists to reuse Las Vegas water to create new habitat for wildlife and to support that productivity in a desert setting. Currently, the riparian areas within the Las Vegas Wash are home for endangered species, such as the desert tortoise, which throughout the Southwest has dictated many land use decisions, and are also suspected to provide habitat for the southwestern willow flycatcher.

One lesson of this project is the demonstration that there is a criti-cal need for obtaining solid biological data and understanding to improve cooperation with the regulatory process. Assessments of several of the endangered species, as for example the flycatcher, actually required five seasonal surveys. These data, in turn, have added to the body of knowledge about the species and will thus be useful for helping flycatchers elsewhere. The surveys have shown that several species were not currently breeding residents of the Wash but that they could be if habitat conditions were improved. During the inventory stage of the planning process, the team came to realize that the capability of the Wash to sustain such wildlife was

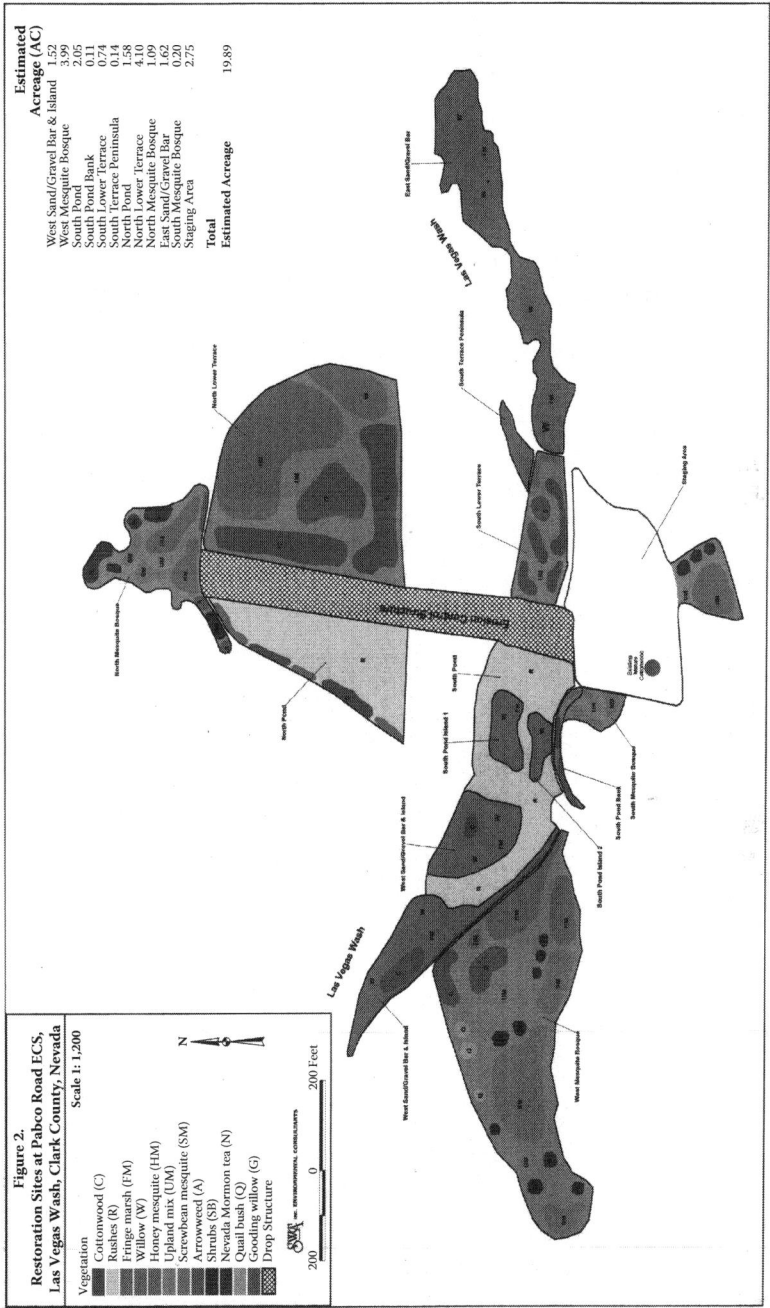

Figure 5.36 Native planting plan in wetland and riparian areas created behind the Pabco Road erosion-control structure.

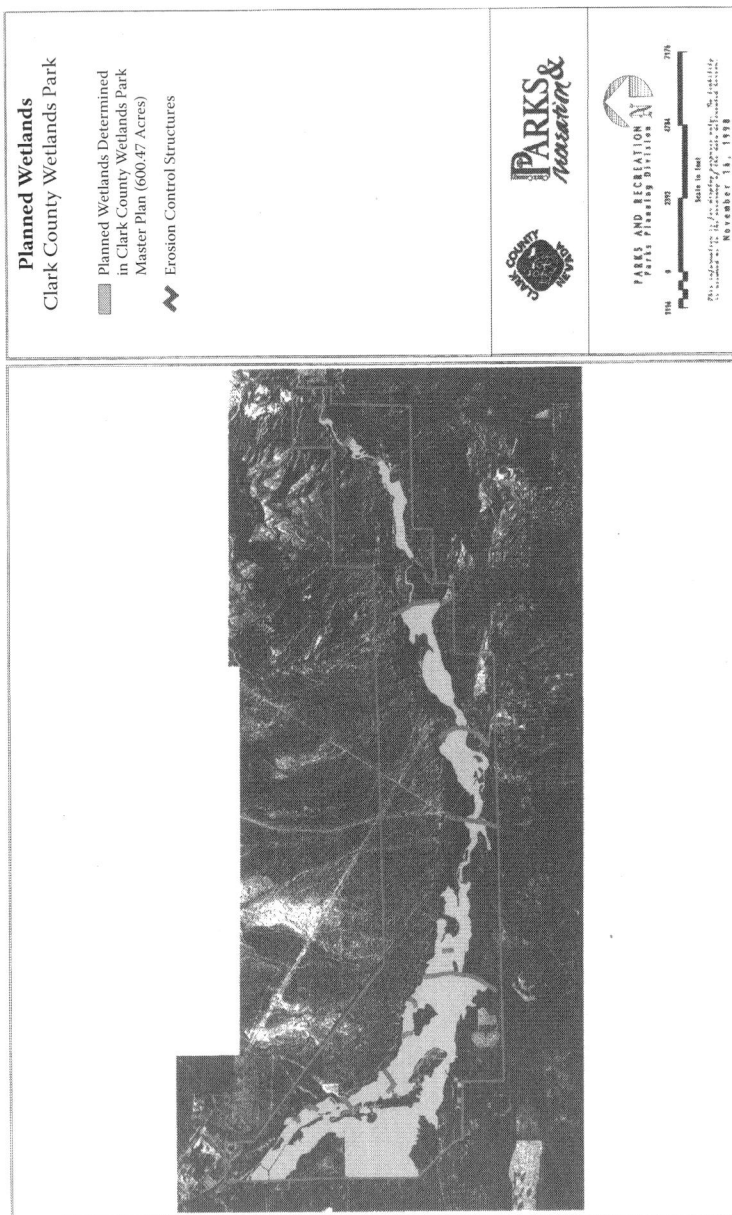

Planned Wetlands
Clark County Wetlands Park

▨ Planned Wetlands Determined in Clark County Wetlands Park Master Plan (600.47 Acres)

◢ Erosion Control Structures

Figure 5.37 Planned creation of restored wetlands in the Las Vegas Wash channel.

far in excess of what it currently supported. Once restoration is complete, the outlook for a biologically diverse wetland park should be good.

There were, however, many challenges (not all of them physical constraints) within the planning process. Planning the site rehabilitation, for example, required extensive regulatory coordination to meld the Clean Water Act and the Endangered Species Act. The willow flycatcher was listed as an endangered species in 1997, and because some ancillary observations suggested it might be present in the Wash, this delayed building the first erosion-control structure for 3 years. The U.S. Fish and Wildlife Service insisted that good data needed to be collected to guarantee that the erosion-control structure would not adversely affect these birds. Despite arguing for the need to build these structures as quickly as possible, the Clark County flood control and parks and recreation departments, not wanting to make waves, waited to obtain flycatcher data from two spring surveys. This delay exacted serious consequences in that during that time, catastrophic flood events destroyed many acres of wetlands. In the end, the irony was that much more wetland habitat was lost through erosion than would have occurred through constructing the first erosion-control structure. In other words, the environmental regulatory process was operating at cross-purposes to the mission of ecological restoration.

Implementation and future plans

As described in the master plan, the Clark County Wetlands Park will be developed over a period from 2000 to 2020. Phasing for the Park was established based on a set of criteria for stabilizing environmental degradation at the same time as increasing public value. Emphasis at the start targeted key elements to provide the foundation for beginning Phase One. Most importantly, a cardinal need was recognized to change the image, reputation, and public perception of the area from the current abuse by a few to a quality wetlands park for all. The essential concept here was to get people to begin to regard the area as a "park." This was recognized to be accomplished only if the following were undertaken: debris cleaned up and all dumping halted, the place becoming a safe public recreation area, the streambank being immediately stabilized to prevent further decreases in the wetland area, and site improvements that were sustainable from an operations and maintenance aspect.

For Phase One, the major focus has been on transforming the image and current misuse of the area into a special place for recreation (Figure 5.38). Funding of approximately $15 million set in place the foundation of the future Park by creating a system of interpretive kiosks and signage, a network of improved and new trails with special viewpoints, construction of the first erosion-control structures, and building of the Visitor Center in a restored surrounding landscape that can be used for

Tabel 4-1: Phasing Criteria Matrix

Legend:
O = Does not Meet Criteria
◑ = Partially Meets Criteria
● = Achieves Goal
n/a = Not applicable

	Transforms Image of the Park	Usable Amenity/Attraction	Cost/Feasibility	Leverage Public/Private Investment	Not Dependent on Erosion Control Structures	Opportunity for Community Participation	Environmental Enhancement	Fits with Construction Phasing
Phase One (Years 1–3)								
Erosion Control Structures (Priority 1)	◑	◑	◑	O	n/a	O	●	●
Visitor Center Site Trailhead	◑	●	●	●	◑	●	◑	●
D-14 Interpretive Area	●	●	●	●	◑	●	●	●
Trails and Trail Amenities	●	●	●	●	◑	●	◑	●
Duck Creek Picnic Area	●	●	●	●	O	●	◑	●
Maintenance/Administration Endowment	●	●	●	◑	●	●	●	●
Phase Two (Years 4–7)								
Erosion Control Structures (priority 2 and 3)	◑	◑	◑	O	n/a	O	●	●
Picnic Pond Vegetation Enhancement	●	●	◑	●	O	●	●	●
Interpretive Facilities at Picnic Pond	●	●	●	●	O	●	●	●
Interior Loop Trails (11,000 ft)	●	●	●	●	◑	●	◑	●
Pabco Trailhead	●	●	●	◑	O	●	◑	●
Henderson Tril Link (3,000 ft)	◑	●	●	O	●	●	O	●
Visitor/Interpretive Center	●	●	◑	●	O	●	◑	●
Phase Three (Years 8–20)								
Trail Expansion to the East (18,000 ft)	●	●	●	◑	●	●	◑	●
Trailhead at Lake Las Veges	●	●	◑	◑	◑	●	◑	●
Upgrade Erosion Control Structures	◑	O	●	O	n/a	O	●	●
Scenic Road	●	●	◑	O	O	O	O	●

Figure 5.38 Matrix of planning criteria for three phases of park development.

establishing an educational program. In Phase Two, the Visitor Center is to be completed and staffed, most of the erosion-control structures constructed, and the trail system expanded. Costs will depend upon several options selected for Phase One. The development of Phase Three completes the erosion-control improvements, finishes the trail system and links it to other public lands, and constructs the scenic drive along the

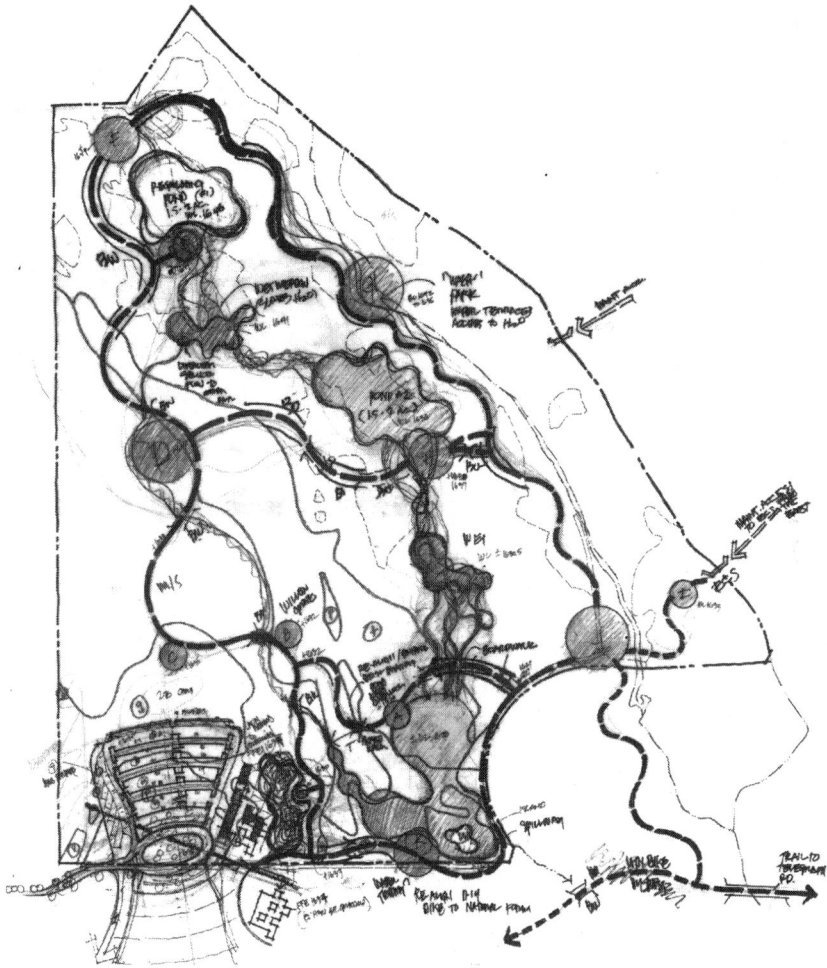

Figure 5.39 Schematic plan for the Nature Preserve showing extensive wash-wetland system.

southern edge of the Park. Enhancements to picnic and viewing areas, interpretive graphic signage, and an eastern trailhead will complete the Wetlands Park. The total estimated cost for all these improvements is in the neighborhood of about $100 million.

The Nature Preserve, comprising about 120 acres in the northwestern section of the Park (Figure 5.39), is to be the first major implementation project and will serve as the heart of the 3,000-acre Wetlands Park. An entry sequence is designed to serve as a decompression zone where visitors transition from the hustle and bustle of urban life to a calming area that splits into a series of trails and loops that offer different experiences as they gradually submerge visitors into the environment of the wetlands. This will be the site of the Visitor Center, an amphitheater for about fifty people, and an overlook to the Wash that is located on an island. Just to emphasize how dynamic the system is, from when the Nature Preserve was first planned, the Wash has shifted its course considerably. This then becomes an enormous challenge: how to design around a system that changes with every storm event? The intended introduction of a major water element that will be interior to the site and arise from channels to the Wash will allow for the opportunity to create new wetlands and open-water areas on the perimeter of the riparian habitat. Two covered shelters will be constructed, the first a bird-viewing blind at the upper water body, and the second a shade structure for gathering and waiting at the bus pickup site. In the latter case, the sloped roof was inspired by the movement of water and is designed to offer respite from different angles of the hot desert sun at variable times of the year.

The Nature Preserve will therefore provide opportunities for wild-life viewing from observation blinds and trails and is designed to serve as a classroom for future generations (Figure 5.40). Instructing about the rehabilitation of nearby wetland plants and riparian trees, the Nature Preserve will be the high-visibility showcase through which to educate about the special, seemingly paradoxical, nature of this location as one of the few desert wetlands in the world. Much attention will be directed toward presenting the strategy for the overall restoration project of the entire Wash. A program of outdoor education is planned for students from nearby schools. By interacting with the Park on a personal level, students will gain an invaluable perspective for their studies as well as a sense of stewardship for both the Park and the greater Las Vegas Valley.

The Visitor Center (Figure 5.41) is intended to be a location where tour groups can get an introduction to the Wetlands Park before they head out on the trail system. Use was made of visual orientation cues to create a sense of place, in particular a connectedness between the adjacent habitat and the distant mountains. In the original plans, a series of interpretive galleries were to be constructed as a set of 20 square-foot modules that can be added or subtracted as needed, depending on developing program needs (Figure 5.42). These modules are designed to (a) "float" 3 feet above the ground, allowing the created wetlands to flow gently underneath year-round, and (b) remain undamaged during extreme flood events when the Wash overflows into its floodplain. The all-glass module walls will allow visitors to be in constant visual contact with the wetlands and mountains. The 8-foot rolling doors will be placed in exterior modules so they can be opened on good weather days. Just like a Japanese garden temple, this will have the effect of complementing the passive solar ventilation system at the same time as further drawing the wetland environment into the galleries. The module roofs would be

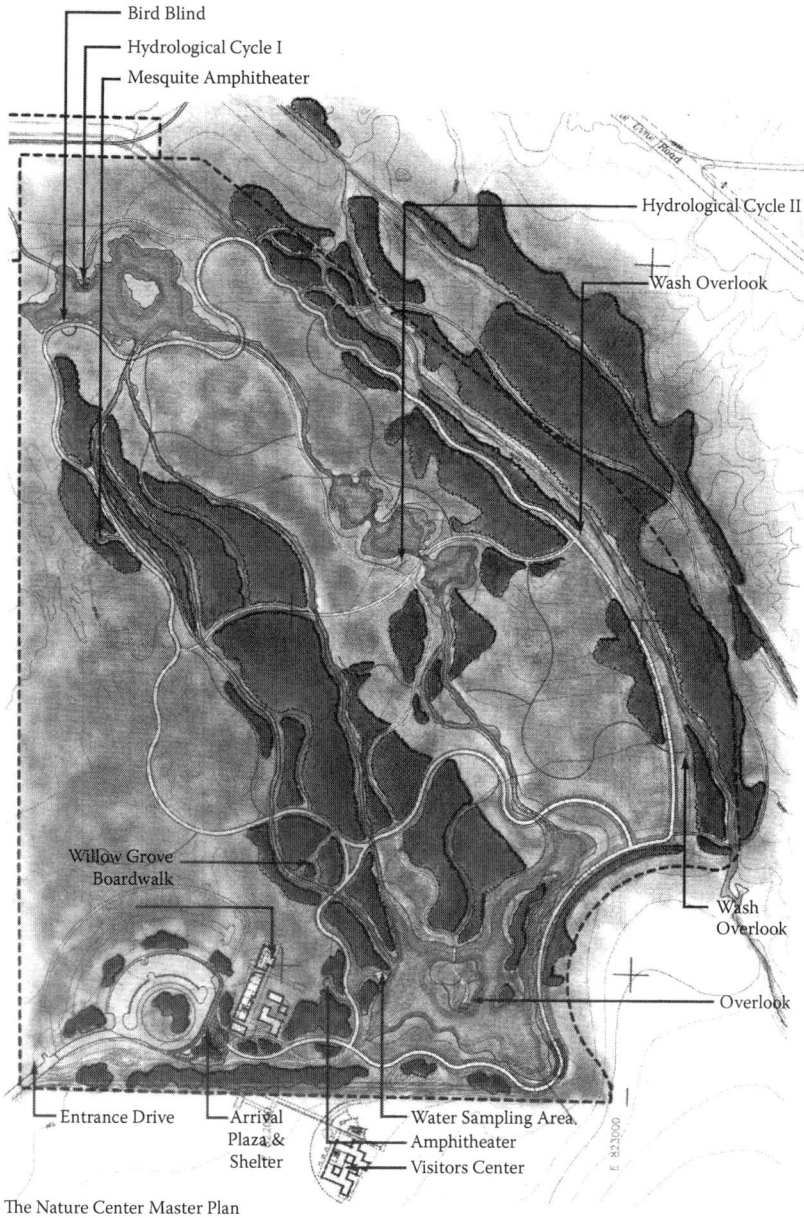

Bird Blind

Hydrological Cycle I

Mesquite Amphitheater

Hydrological Cycle II

Wash Overlook

Willow Grove
Boardwalk

Wash
Overlook

Overlook

Entrance Drive

Arrival
Plaza &
Shelter

Water Sampling Area

Amphitheater

Visitors Center

The Nature Center Master Plan

Figure 5.40 Master plan for the Nature Preserve.

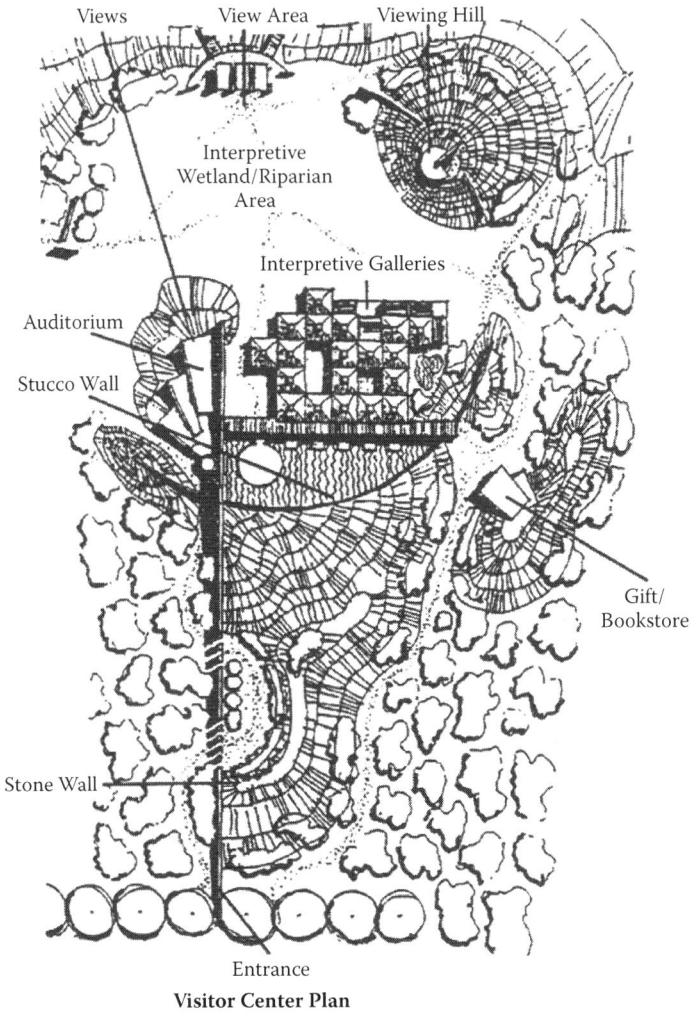

Views   View Area   Viewing Hill

Interpretive
Wetland/Riparian
Area

Interpretive Galleries

Auditorium

Stucco Wall

Gift/
Bookstore

Stone Wall

Entrance
Visitor Center Plan

Figure 5.41 Site plan for the Visitor Center.

made of metal, which will be able to be seen softly shimmering from distant trails of the Nature Preserve, thereby providing a visual anchor from which to leave the "home base" and venture forth to explore the more remote corners of the Park.

Designs for the Visitor Center build upon the earlier visions. The end goal is to design an interpretive center that will fit integrally into the landscape by blurring the edge between outdoors and indoors through assuming a long, low profile and subtle colors to blend in with the desert setting (Figure 5.43). Significantly, and in line with the environmental

Visitor Center Model Simulation

Interpretive Wall and Path Simulation   Boardwalk and Galleries Simulation

Research, Administrative Building Simulation

Figure 5.42 Early simulations of the Nature Center.

Figure 5.43 Final simulation of the Visitor Center.

Northwest Elevation
Entrance

Figure 5.44 Simulation of Visitor Center raised on pylons.

Where curiosity may be both piqued and satisfied. Standing on the deck of the central courtyard, the kiosk's shape echoes that of the circular courtyard which opens to the sky above and the wetlands blows.

To nature preserve trails   *Wheelchair-accessible ramp*

Figure 5.45 Simulation of approach to Visitor Center.

sensitivity of the area, a goal of achieving a platinum Leadership in Energy and Environmental Design (LEED) certification has been set for the building. The strategies to be employed include such elements as creating shaded zones to minimize heat island effects, rain harvesting and graywater reuse for irrigation, use of recycled building materials, and use of renewable energy sources. Much thought has been devoted to employing passive solar techniques, such as the correct window orientation for primary interior light, and type of exterior panels to reduce seasonal heat and glare. To encourage immersion into the experience of the Wetlands Park, the walkway from the parking lot to the building will pass over a small section of the trail system as it curves alongside interpretive wetlands in the Wash. As originally conceived, the entire structure will be elevated on pylons to ensure safety from flash floods (Figure 5.44).

Visitors will approach the building by ascending a crescent-shaped ramp (Figure 5.45) at the top of which will be a cylindrical information kiosk whose shape echoes that of a circular central courtyard which

Southeast Corner   *Auditorium, exhibit hall, observation deck*

Auditorium   *Windows look toward Calico Ridge and Lake Mead*

Figure 5.46 Simulations of auditorium at the Visitor Center.

opens from the wetlands below to the sky above and offers views of the surrounding Wash scenery. The darkened, 200-seat auditorium will show a film introducing visitors to the Wetlands Park (Figure 5.46). In a thespian flourish, as the video concludes the projection screens will rise, curtains part, sunlight will flood in, presenting a magnificent view of the wetlands and surrounding mountain ranges. Exhibits—illuminated by natural sunlight as much as possible (Figure 5.47)—will expand on information presented in the auditorium and focus on three primary topics: the history of human occupation of the wetlands, the geologic history of the Wash, and the ecology of the wetlands. Art, living displays, dioramas, giant models of animals, and an "armchair tour' area with real-time video displays will be used as education tools.

Trail circuitry was based on travel times in loops of 10, 30, 60, and more minutes, which provided multiple ways to engage the site. The trails were designed in a hierarchy (Figure 5.48): Larger primary trails are 8 feet in width and made of concrete with about 2 feet of crusher fines along edges and are reinforced to enable maintenance trucks to access the site; secondary trails are composed of 6 feet of crusher fines; and tertiary trails are simple footpaths, 1.5 feet wide and formed of dirt. The Clark County

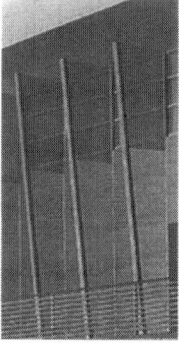

south shade fins

Exterior shade fins screen the Armchair Tour's north wall of windows

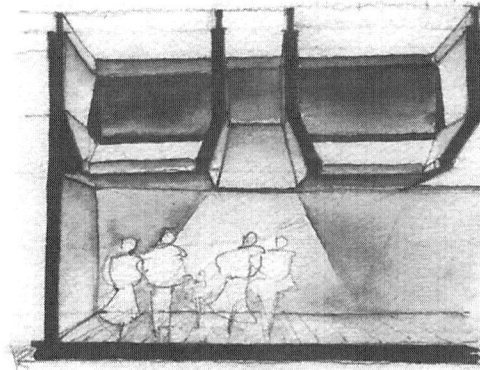

... provides energy-efficient interior lighting

channeled and diffused sunlight ...

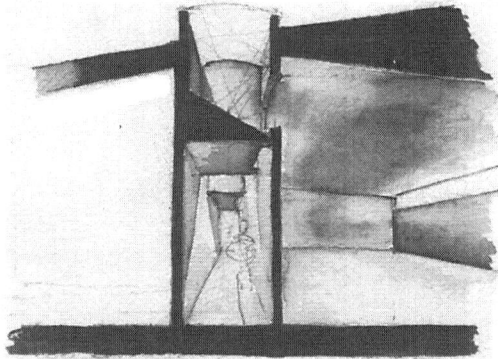

Figure 5.47 Architectural sketches showing how natural light will enter the exhibit areas.

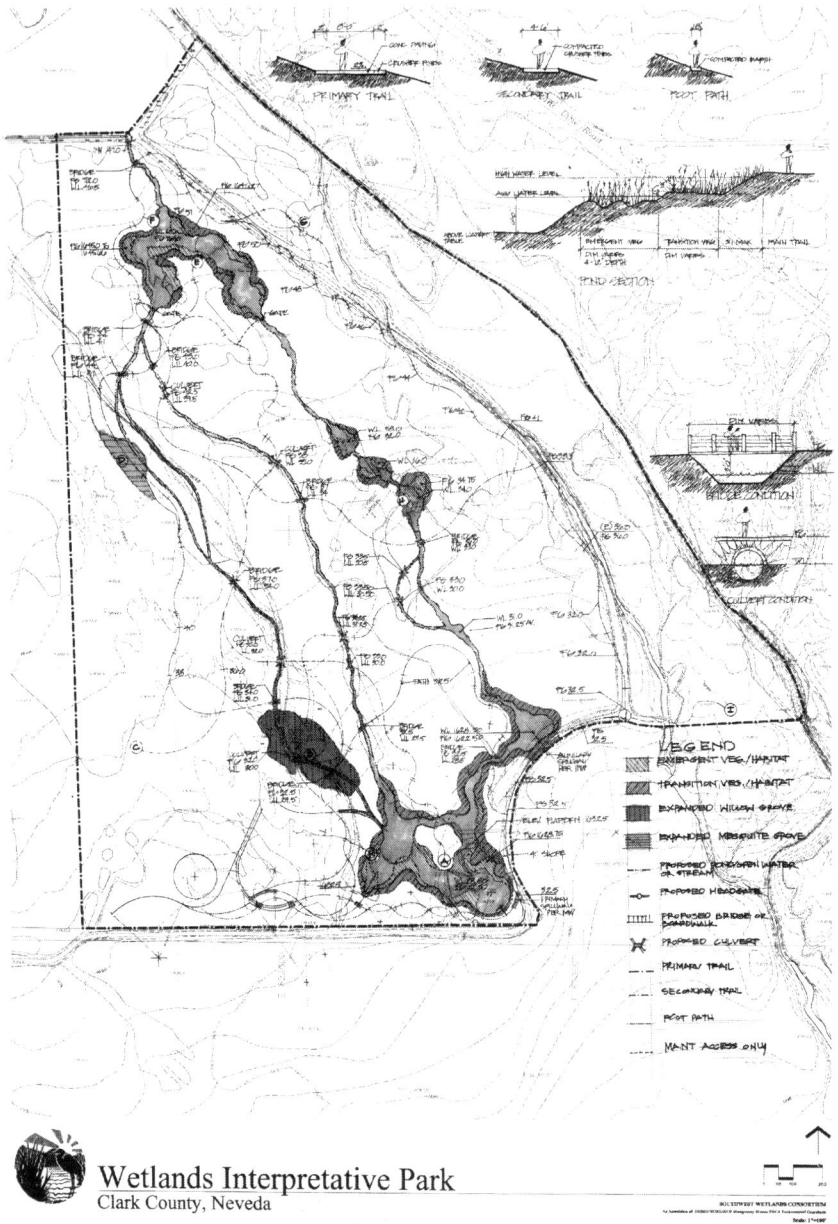

Figure 5.48 Network and hierarchy of trails at the Nature Preserve.

Legend

Primary Trail

Secondary Trail

Footpath

Trails Plan

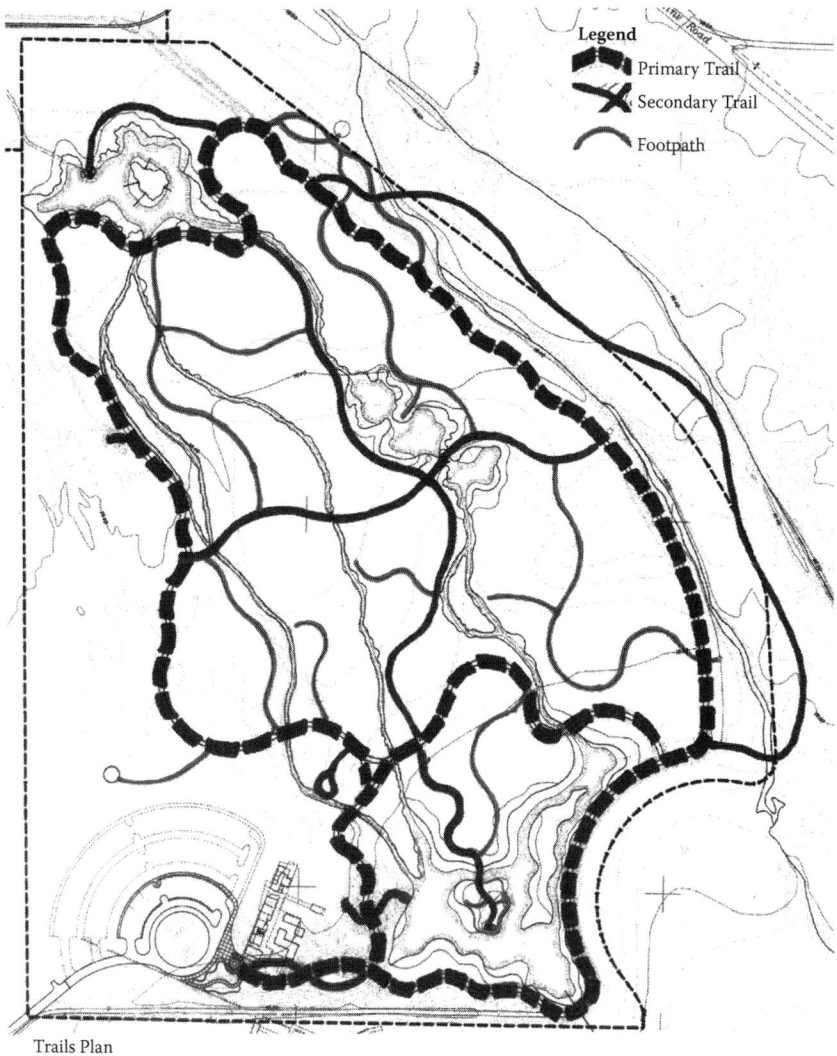

Figure 5.48 (continued)

Parks and Recreation Department was able to construct a large portion of the trail network in less than half a day through the donation of time and labor of more than a thousand American Express employees participating in a company-sponsored volunteer program.

The imaginative use of such volunteer labor has been a compelling lesson of successful project implementation gleamed from the Clark County Wetlands Park. For example, the information kiosk was one of a series of structures built at the head of the new trail system in the Nature Preserve by volunteer labor in a single day. As well as providing interpretive information, such structures are essential for shade and shelter in such a desert climate. Since all the structures were built by unskilled or semiskilled volunteers, the design relied upon a very simple off-the-shelf kit of materials costing no more than $1500 apiece and composed of easily assembled, bolted-together material. Additionally, the woven metal walls and roof of the information kiosk call to mind the construction techniques used by the Valley's native inhabitants.

As an information brochure mentions: "If plants and animals are the embodiment of the Nature Center, water is the soul." As such, water has been the guiding spirit in the landscape design of the Nature Preserve and Center. Three sources of water supply the site, all of which will be incorporated into the educational program. The Wash itself runs along the eastern side of the site and is, of course, fed by treated wastewater that, ironically, is used to improve the quality of the stormwater from the city. More than anything, this promises to be a sobering lesson about the quality of urban runoff—imagine the shock to learn that treated sewage is actually cleaner than the rainwater coming off the streets! Stormwater also enters the site from both the northwest and southwest directly through drainage ditches that open into large wetland ponds created to improve water quality before it is released into the Wash.

The challenge for projects with long implementation schedules is to maintain the vision from start to finish. Such visions are often embedded in individuals who were instrumental in creating the initial steps. Over time, though, the individuals move on, the political climate changes, and implementation becomes compromised. So what has to happen is that an entity needs to be created at the very start to function as a sort of "vision keeper."

In 1998, the South Nevada Water Authority, recognizing that water quality issues related to the Wash are too complex (both scientifically and sociologically) to be the responsibility of any single public entity, established the 28-member Las Vegas Wash Coordination Committee, comprised of local, state and federal agencies, public citizens, business leaders, and representatives of various environmental groups. The Committee's mandate was to research and analyze issues relating to the Wash and to develop the Comprehensive Adaptive Management Plan that will recommend solutions to the problems.

Where does the Clark County Wetland Park proceed next? The immediate 5-year plan is very aggressive: 350 acres of wetlands will be added, another 10 miles of trails will be improved, at least 4 to 6 and possibly 10 erosion-control structures will be constructed, the Visitor Education Center and scenic drive will be built, and the remaining 275 acres of privately held land in the Park will be acquired. This work will be stewarded

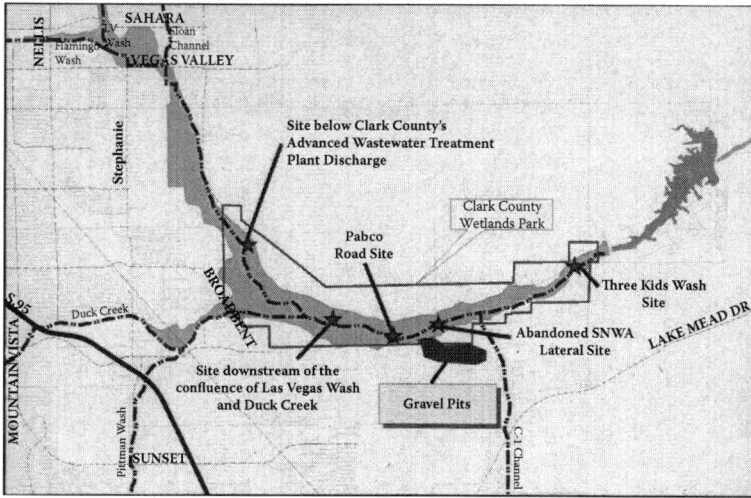

Figure 5.49 Abandoned gravel pits as a possible future location for wetland creation.

by the overseeing Las Vegas Wash Coordination Committee and the Comprehensive Adaptive Management Plan. The plan is intended to function as a tool to help those agencies who have the responsibility and authority over aspects of the Wash. It is to be looked upon as a living document that will be updated periodically to reflect progress and changing conditions of the Wash.

The Adaptive Management Plan comprises the efforts of nine study teams that were established to identify and analyze key issues and concerns in depth from which to generate recommendations and actions: Shallow Ground Water; Erosion and Stormwater; the Wetlands Park; Alternate Discharge; Environmental Resources; Land Use; Jurisdictional and Regulatory; Public Outreach; and Funding.

One topic that was examined by the Erosion and Stormwater Study team concerned the creation of wetlands outside of the Wash channel, in addition to those that will naturally appear in the channel through impounding water behind the erosion-control structures. Given the ongoing erosion forces within the Wash channel, it is unlikely that large areas of wetlands will be able to become established there. The idea would therefore be to capture the stormwater surges in retention wetlands constructed in locations outside the Wash. The study team therefore recommended that an investigation of location feasibility be undertaken to examine and evaluate sites such as abandoned gravel pits for this purpose (Figure 5.49).

The Wetlands Park Study team developed six recommended actions to guide Clark County in implementing their master plan: identify water

PUBLIC OUTREACH PROGRAM TACTICS

Tactic	Purpose	Frequency
News Releases	The public, and even to some extent the media, will be patient with the management plan's development as long as they sense progress is being made.	As needed
"The Current" Newsletter	Stakeholder interviews have indicated that people interested in the Wash and the management plan's development would like to be apprised of Coordination Committee progress through a newsletter.	Quarterly
Water Quality Reports	Water purveyors responsible for distributing a consumer confidence report will include information about efforts being undertaken by the Coordination Committee.	Annual
Lobby Displays	Lobby displays provide an easy mechanism to maximize exposure with limited resources.	Ongoing
Internet Web site	Increasingly popular Internet technology will allow interested citizens access to the most current information available, and provide a forum for feedback and/or discussion of key issues.	Ongoing
Public Scoping Meetings	It is essential that the public have a voice in matters relating to the Wash. Public scoping meetings will allow for an open dialog between the community and Coordination Committee in an effective, efficient manner.	Ongoing
Children's Educational Program	The goal of the Children's Educational Program is to teach children about the significance of the Wash as natural resource, water quality, and the monitoring process.	Ongoing
Media Briefings	Media briefings serve to impart an understanding of the issues at stake and a general knowledge of how implementation of the management plan will affect those issues.	Annual
Stakeholder Awareness Briefings	If the Coordination Committee is to establish a united position on the Wash, it is imperative that employees of participant entities have a general understanding of the Wash and the issues surrounding it.	Annual

Figure 5.50 Tactics for public outreach developed for the Adaptive Management Plan.

PUBLIC OUTREACH PROGRAM TACTICS

Tactic	Purpose	Frequency
Speakers Bureau	Implementation of a speakers bureau will serve as a tool to enable the team to hold an open discourse with members of the community, to determine their concerns and provide timely, accurate information.	Ongoing
Las Vegas Wash & Wetlands Clean-Up	This event creates a positive visual presence, raises public awareness of the Wash, improves the Wash from a pollution standpoint, and helps foster environmental stewardship within the community.	Annual
Water Information Fairs	Water information fairs offer a special opportunity to promote the Wash and the efforts of the Coordination Committee directly with the public through staff interaction, print materials, and displays.	As needed
Las Vegas Wash "Familiarization Tours"	Key constituents should be led on tours of the Wash early in the process to establish "before and after" visuals that will help underscore the project's urgency and develop a baseline perspective for its progress.	Ongoing
Interested Regional Stakeholder Outreach	Because water from the Colorado River is a shared resource, there is considerable interest outside southern Nevada that requires regional outreach efforts.	As needed

Figure 5.50 (continued)

resources needed to maintain the Park; develop long-term monitoring plans; develop a long-term operations and maintenance plan; ensure implementation of mitigation measures; and identify funding needs. Although all these issues were generally outlined in the 1995 Master Plan, in the 2001 Adaptive Management Plan, they are explored and expanded upon and particular approaches identified with specific actions targeted.

The recommendation and development of the Coordination Committee website (www.lvwash.org) by the Public Outreach Study team has been a major step in providing access to current information on issues such as erosion, water quality, habitat restoration, project updates, materials for teachers, and almost a thousand photos and movies. Other recommendations from this group address news releases, the newsletter, water quality reports, visual displays, public scoping meetings, children's education programs, media briefings, stakeholder awareness raising, and the cleanup days (Figure 5.50).

"Adaptive management" became the buzzword for managing the Wetlands Park. In other words, any planned activity that is not in the ground has to be flexible and able to adjust to the dynamic nature of the Park. For example, the Pabco Road erosion-control structure was redesigned three times during the period of the construction delay. In the Clark County Wetlands Park, the only constant is therefore change. All activities must be comprehensive and able to be integrated into the other planning efforts. Now that formal implementation has begun, it is critical to maintain the momentum and the vision. Being adaptive means the ability to work around a property owners who may be unwilling to sell their property at that particular time and/or be responsive to other environmental issues as they arise. "Expect the unexpected" is the mantra for successful adaptive environmental management.

Maintaining project momentum is largely a function of political will. Along with acquiring the spatial investment for site expansion, resources are also required for ongoing day-to-day operation, maintenance and staffing. Complementing the existing political will is the need for generating local community support (Figure 5.51), which cannot be understated. There is a huge volunteer effort in the Wetlands Park, with more than 20,000 volunteer hours spent each year. In one example, 10,000 trees and shrubs were planted by volunteers on six sites within the Wash. A growing number of local firms and public agencies are supporting and sponsoring projects. The school district, the university, and federal and state agencies are active participants in the continual development of the Park. As a result, sections of the Park have been set up as outdoor science laboratories.

A healthy future of the Park depends on the collaborative effort involving elected officials, businesses, local communities, and educational institutions. For example, all truly sustainable, and therefore successful, environmental restoration projects are as much about restoring degraded human–nature relationships as they are about simply repairing degraded physical landscapes. As such, mobilizing public interest and support was deemed essential right from the very start of planning, and with this in mind, future plans will specifically address the countless opportunities that are expected to develop for volunteers to assist with the ongoing management of the park. Some volunteers may wish to help provide interpretive programs, others may want to work on special projects such as trail construction, revegetation, or cleanup. By involving people in the direct functioning of the Park, a feeling of ownership is nurtured such that the likelihood of the site being "orphaned" becomes minimal.

Are you the person who can help us?

Imagine a great, green, lush oasis in the middle of the Mohave Desert. A place of water that is alive with wildlife and plants, such as heron nesting in willows and waterfowl dabbling amongst cattails and reeds. Imagine what such a place—sitting on our communities' very doorstep—would mean, to residents and visitors. Just such a place is Las Vegas Wash, a desert wetland in Clark County between the city of Las Vegas and Lake Mead. Originally a grassy wash, it became especially verdant and oasis-like as it started to be fed by waste waters leaving the valley. In recent years, however, it has begun to decline because too great a flow of water, largely from urban runoff, has badly eroded it. A master plan for the Clark County Wetlands Park has been approved which will guide the restoration, protection, and enhancement of the Wash. This environment's potential is finally being realized. Read on—to find out how you can become involved...

Clark County Parks and Recreation
Parks Planning Division
2601 East Sunset Road
Las Vegas, Nevada 89120

Clark County Board of Commissioners
Yvonne Atkinson Gates, Chair
Lorraine T. Hunt, Vice-Chair
Erin Kenny • Mary J. Kincaid • Lance M. Malone • Myrna Williams • Bruce L. Woodbury

Dale W. Askew, County Manager
Bonnie Rinaldi, Assistant County Manager
Mike Alastuey, Assistant County Manager
Glenn Trowbridge, Director of Parks and Recreation

Clark County does not discriminate on the basis of race, color, national origin, sex, religion, or disability in employment or the provision of services.

Printed on recycled paper

Figure 5.51 Information brochure soliciting for volunteers for the Wetlands Park.

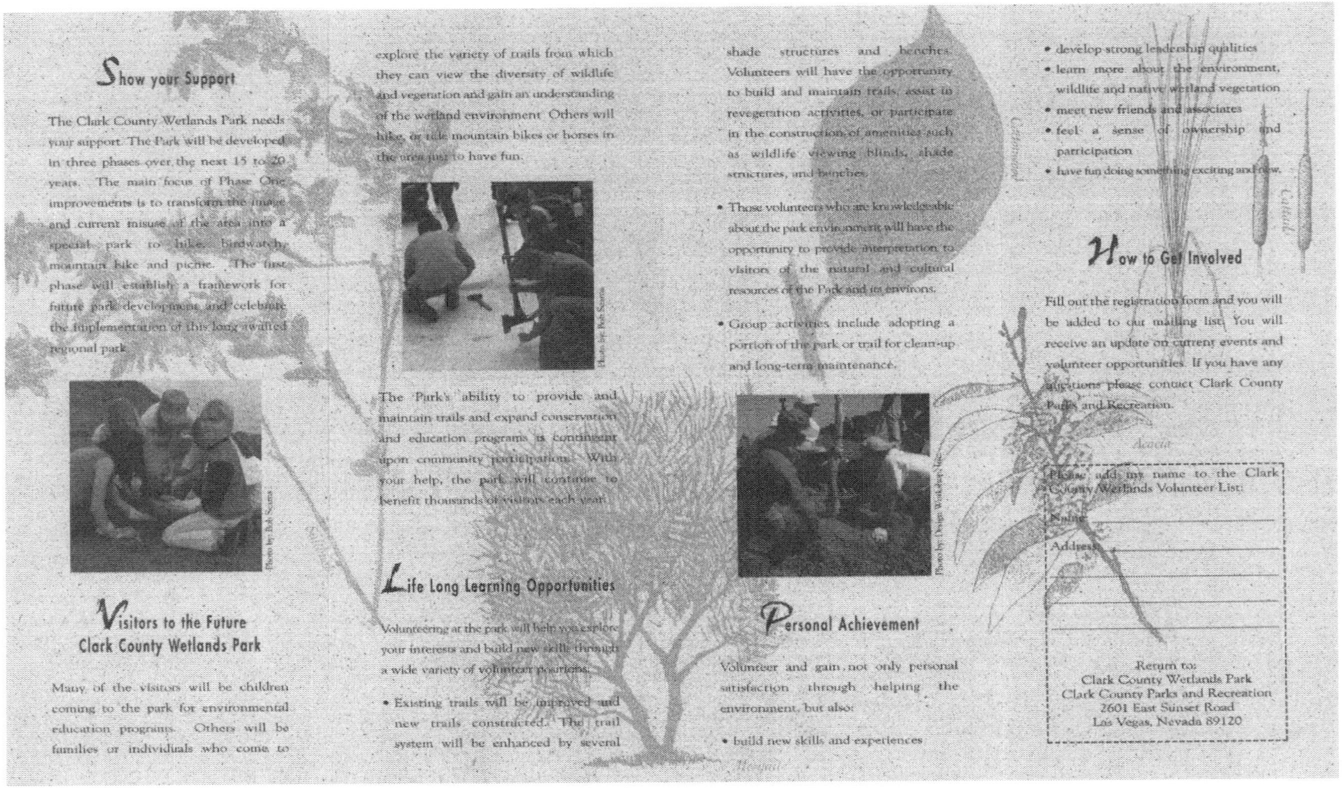

Show your Support

The Clark County Wetlands Park needs your support. The Park will be developed in three phases over the next 15 to 20 years. The main focus of Phase One improvements is to transform the image and current misuse of the area into a special park to hike, birdwatch, mountain bike and picnic. The first phase will establish a framework for future park development and celebrate the implementation of this long-awaited regional park.

Photo by: Bob Sears

Visitors to the Future Clark County Wetlands Park

Many of the visitors will be children coming to the park for environmental education programs. Others will be families or individuals who come to explore the variety of trails from which they can view the diversity of wildlife and vegetation and gain an understanding of the wetland environment. Others will hike, or ride mountain bikes or horses in the area just to have fun.

Photo by: Bob Sears

The Park's ability to provide and maintain trails and expand conservation and education programs is contingent upon community participation. With your help, the park will continue to benefit thousands of visitors each year.

Life Long Learning Opportunities

Volunteering at the park will help you explore your interests and build new skills through a wide variety of volunteer positions.

- Existing trails will be improved and new trails constructed. The trail system will be enhanced by several shade structures and benches. Volunteers will have the opportunity to build and maintain trails, assist in revegetation activities, or participate in the construction of amenities such as wildlife viewing blinds, shade structures, and benches.
- Those volunteers who are knowledgeable about the park environment will have the opportunity to provide interpretation to visitors of the natural and cultural resources of the Park and its environs.
- Group activities include adopting a portion of the park or trail for clean-up and long-term maintenance.

Photo by: Design Workshop, Inc.

Personal Achievement

Volunteer and gain not only personal satisfaction through helping the environment, but also:

- build new skills and experiences
- develop strong leadership qualities
- learn more about the environment, wildlife and native wetland vegetation
- meet new friends and associates
- feel a sense of ownership and participation
- have fun doing something exciting and new.

How to Get Involved

Fill out the registration form and you will be added to our mailing list. You will receive an update on current events and volunteer opportunities. If you have any questions please contact Clark County Parks and Recreation.

Please add my name to the Clark County Wetlands Volunteer List:

Name: ______________________

Address: ______________________

Return to:
Clark County Wetlands Park
Clark County Parks and Recreation
2601 East Sunset Road
Las Vegas, Nevada 89120

Figure 5.51 (continued)

Sources and references

Ahlstrom, R. 2005. *Desert oasis: The prehistory of Clark County Wetlands Park.* Henderson, Nevada. HRA Inc., Conserv. Author.

Anon. 1987. *Las Vegas Wash revegetation study.* Bureaus of the Interior and of Reclamation.

Anon. 1995. *Las Vegas Wash, Clark County Wetlands park planning process.* Southwest Wetlands Consortium.

Anon. 1995. *Las Vegas Wash, Clark County Wetlands Park master plan.* Southwest Wetlands Consortium.

Anon. 1997. *Las Vegas Valley Watershed wastewater needs assessment study.* Montgomery Watson.

Anon. 1997. *Clark County Wetlands Park brochure.* Clark County Parks and Recreation.

Anon. 1997. *Framework for comprehensive management of the Las Vegas Wash.* Southern Nevada Water Authority.

Anon. 1997. *Forward planning for wastewater services executive building.* City of Henderson.

Anon. 1997. *The City of Henderson water reclamation facility and bird viewing preserve. A dual purpose facility.* City of Henderson.

Anon. 1999. *Las Vegas Wash engineering workshop.* Las Vegas Wash Project Coordination Team.

Anon. 2000. *The nature center at Clark County Wetlands Park.* Clark County Parks and Recreation.

Anon. 2000. *Las Vegas Wash erosion mitigation project.* Kennedy/Jenks/Chilton and Associates.

Anon. 2000. *Las Vegas Wash comprehensive adaptive management plan.* Las Vegas Wash Project Coordination Team.

Anon. 2001. *Lake Mead National Recreation Area: Las Vegas Wash stabilization project environmental assessment.* United States Dept. of the Interior, National Park Service.

Anon. 2001. *The Nature Preserve Visitor and Education Center, Clark County Wetlands Park.* Clark County Parks and Recreation.

Anon. 2008. lvwash.org

Brothers, K. 2002. Water wars: From competition to cooperation. *Urban Land Magazine,* Sept.:110–13.

Christensen, J. 1994. Las Vegas wheels and deals for Colorado River water. *High Country News,* 21 February.

Davis, J. C. 2000. Saving the Las Vegas Wash. *Erosion Control Magazine.* Jan/Feb.

Davis, M. 2007. "The Strip versus nature." In *Metropolis now!: Urban cultures in global cities,* ed. R. K. Biswas. New York: Springer.

France, R. L. 2001. (Stormwater) leaving Las Vegas. *Landscape Architecture Magazine,* 8:38–42.

France, R. L. 2003. *Wetland design: Principles and practices for landscape architects and land use planners.* New York: W.W. Norton.

France, R. L. 2010. "Integrated restoration and adaptive management of the Las Vegas Wash," in *Restorative redevelopment of devastated ecocultural landscapes,* ed. R. L. France, 117–124. Boca Raton, FL: CRC Press.

France, R. L. 2010. "Planning and development of a desert wetland park in the United States." In *Restorative redevelopment of devastated ecocultural landscapes,* ed. R. L. France, 219–226. Boca Raton, FL: CRC Press.

France, R. L., ed. 2005. *Facilitating watershed management: Fostering awareness and stewardship.* Lanham, MD: Rowman & Littlefield.

France, R. L., ed. 2007. *Healing natures, repairing relationships: New perspectives in restoring ecological spaces and consciousness.* Winnipeg: Green Frigate Books.

Fred Phillips Consult. 2002. *Clark County Wetlands Park trail corridor and guidelines plan.*

Hester, G., and K. Grear. 2002. Erosion control lessons being learned on the Las Vegas Wash. *Land and Water Magazine,* 46:31–39.

Hong, Z. 2007. *A look into Las Vegas's water management past, present, and future.* Watershed management, restoration and design course, Harvard University.

Johnson, J. 1999. *Estimation of stormwater flows in Las Vegas Wash, Nevada and potential stormwater capture.* Southern Nevada Water Authority.

Kao, C. 2001. *The Clark County Wetlands Park as a political landscape.* Penny White Award Project. Harvard Graduate School of Design.

Lin, T-Y. 2007. *Learning from the restoration design in Las Vegas.* Watershed management, restoration and design course. Harvard University.

Schumacher, G., and R. Bryan. 2004. *Sun, sin and suburbia: An essential history of modern Las Vegas.* Las Vegas: Stephens Press.

Selby, S. 1999. *Colorado River water return flow credits—An important component of southern Nevada's current water resources.* Southern Nevada Water Authority 35p.

Venture, R., D. Scott Brown, and S. Izenour. 1977. *Learning from Las Vegas: The forgotten symbolism of architectural form.* Cambridge, MA: MIT Press.

Weissenstein, M. 2001. The water empress of Vegas. *High Country News,* April 9.

Appendix:
City of Henderson Water Reclamation Facility and Bird Viewing Preserve

chapter 6

Las Vegas case study questions

Answers by Mark Raming, Vicki Scharnhorst, Becky Zimmerman, and Jeff Harris

How exactly were the wetlands lost? Was it because they were scoured away by too much water during extreme storms? Or was it because they were stranded high and dry with too little water because most is now trapped deep down into the eroded wash channel and no longer spreads out laterally in the former floodplain?

Both these processes were active players in the wetland loss together with an overall deepening water table.

Given that the wetlands are not really "natural" but are rather the product of wastewater from Las Vegas, what is the wisdom in "restoring" such artificial systems in a landscape where they are not really meant to be? In this respect, how have the local naturalist groups reacted to the project?

The word "natural" has lost much meaning in general and with reference to the Las Vegas regional environment in particular. This project has enabled an opportunity to recreate desert riparian habitat, a type of ecosystem that has been seriously depleted throughout the Southwest because of intensive water management for human uses. Local naturalists have generally reacted well to that awareness.

Certainly one major challenge must be finding ways to connect people and attract volunteers in Las Vegas (as opposed, for example, in London), a city that is growing so rapidly and in which the residents are for the most part recent arrivals with no vested history in the local landscape.

Not really. The stable community is always there and there is a growing influx population of seniors who are always eager to volunteer. For example, the temporary visitor center is currently open from 10:00 A.M. to 4:00 P.M., 7 days a week, and is entirely staffed by volunteers. There is also a nearby community of homes exclusively inhabited by horse owners. Because sites for equestrian activity are in increasingly short supply as suburbs become urbanized, these individuals are very protective of the green spaces offered by the Park. As a result, they have been approached to form a sort of supervisory "free militia cavalry" in return for building them a horse trail. And once a year, there is a general Wash cleanup, which in the first year had close to five-thousand people show up. In the second year, as many as six-thousand people participated, and for the third year the event was cancelled because all the trash had been picked up.

Many of the recreation opportunities to be created in the new Park appear to be passive, in line with the ecological mandate of the project. Is there a fear that the active recreation that until recently has taken place here, such as dirt bike riding, etc., will simply be pushed away elsewhere to perhaps more pristine and fragile true desert environments?

To a certain extent, the true desert environment in the Las Vegas vicinity has already been hammered by off-road vehicle use. Closure of the relatively small area of the Wash for these opportunities is not expected to have much of a displaced activity effect.

What is the nature of the Rainbow Gardens area in terms of its unique geology? And will this be integrated into the Park?

The Rainbow Gardens are pastel-colored rock strata that have been tilted and exposed. They lie on public land that will be accessible from the Wetland Park.

The Wash is an effluent-dominated stream, and given that water is in short supply in Las Vegas, how does the issue of reclaimed water rights get handled? And how have elements in the system been protected from the very high runoff events?

The two significant water sources for the Wetland Park are the effluent stream and the urban runoff. We have negotiated with the state engineers for 8,000-acre feet of water rights. Never before in the State has such a primary water right for urban runoff been proposed and awarded. And we also have secondary water rights on treated effluent.

With respect to obtaining the use of the reclaimed water, there were some challenges. The area is home to many golf courses that are, however, restricted to use of only about 20 percent of the reclaimed water during the winter months, thereby always ensuring an ample supply of reclaimed water for the Wash as base flow.

In 1983, a flood control bond was passed to create numerous aboveground detention basins. Although these may become public dual-use facilities, it has not been a priority. The good news, however, in terms of water management is that they do catch the high-volume flash floods—predominantly up near the base of the mountains—and those that are constructed have done a reasonable job of attenuating flooding. Also, these structures are designed for the probable maximum storm, considered to be a 500-year event, whereas the structures in the Wash are all sized for the 100-year storm.

The whole "return-flow credits" system of water trading appears to be an impediment to Las Vegas facing reality and coming to grips with the fact that it is situated in a desert and therefore should do a better job at conserving water. Is there any way around this?

The only solutions are political ones; if Las Vegas doesn't take steps in this direction, be assured that downstream water users such as Phoenix and Las Angeles will.

Henderson, a suburb of Las Vegas, apparently leads the country in reclaimed water use and has transformed their treatment ponds into a very popular bird-viewing nature preserve that is situated immediately to the south of the Wetlands Park. What, if any, are the relationships, either physical or institutional, between the two organizations?

Both cities participate in the greater Las Vegas Valley municipal district administration activities. Besides physical proximity, there are few integrative links besides the birds that fly back and forth between the various water bodies.

What about the controversy regarding control of the exotic tamarisk in relation to it also supporting rare wildlife? And is there a plan to make a dent in the salt cedar that is there now?

The tamarisk is a challenge, although the 4 years of survey data indicate that there are no flycatchers there at present. The Fish and Wildlife Service is very cooperative in letting us take the tamarisk out and putting the erosion-control structures in to create the oases of restored area that will be intensively managed as revegetated pockets. Many of the supporters of the Wash and Wetland Park wanted to do a wholescale removal of tamarisk over the entire site, but we strongly discouraged this approach because it would be incredibly difficult to implement successfully. Instead, we proposed a staged approach of focusing on 20-acre parcels that are managed by hand along with other controls of beavers, etc., in addition to monitoring the flycatchers' response to all this.

The strategy of developing a few, intensively managed areas and then letting the effects of these spread elsewhere is interesting. Will the educational aspects of this, in other words, the self-healing potential of nature, be developed in the Park's program?

Yes, this is a major lesson arising from the planning and management of the Clark County Wetlands Park.

How many wildlife species exist in the Park and what changes have been noticed as the wetlands become restored?

There are hundreds of species present, though no direct population estimates have been made. That said, anecdotal evidence suggests an increase in the abundances of migrant and resident waterbirds and shorebirds.

The issue raised about how delays caused by adhering to the draconian environmental regulations actually resulted in more damage to the system seems particularly important, not just for this particular project, but also for undertaking reclamation work elsewhere. How much of a limitation is this in the search for, and implementation of, innovative ecological restoration?

It can be very limiting. The Wash faced greater scrutiny because of the massive and quite unnatural aspect of the flood-control structures. The whole issue about what is "natural" and what must be protected has gotten twisted around beyond usefulness from a reparative management point of view.

What about the migration of sediment materials over time?

The impounded areas behind the erosion-control structures where sediment builds up are designed in such a way to allow for dredging to maintain an open-water area. We expect that once all the structures are in place and doing their job, the overall maintenance will drop right off.

Under normal conditions, sediment load remains fairly low. Another issue that is tied to the question of sediment management is the associated pollution load of the sediments in those areas. The absorption rate needs to be examined. The Fish and Wildlife Service was critical and thought that we were creating pollution sinks. We responded that because we did not really know whether this was the case, we wanted to build several structures in which we could monitor whatever is desired in order to find an answer to the question.

Elevated perchlorate concentrations have been found in the groundwater. What is this substance, where does it come from, and how has it affected the design of the erosion-control structures?

Perchlorate is a byproduct of rocket fuel and explosives manufacturing that in this case originated from several plants in Henderson that have been in operation since the Second World War. Erosion-control structures were designed to enable periodic collection and removal of contaminated sediments for treatment.

There appears to be an irony in all this with respect to Lake Mead being the drinking water supply for Las Vegas. Do the citizens appreciate the fact that everything that gets washed off their streets and down their toilets will, after passage through the Wash and Lake Mead, get recycled back to them?

Not really; most naively believe that the water comes from the mountains.

Have there been any attempts to introduce water-sensitive low-impact development planning for the new sprawl communities being built around Las Vegas, many of which are actually surrounded by moats that dump their water straight into the Wash?

Yes, actually Las Vegas has become much more water conscious over the last two decades. Although the population has grown considerably, the per-capita water consumption has actually dropped by about 70 percent.

What do the developers think about all this potential real estate being locked up in a protected park? Are some attempting to redesign their communities to take advantage of being adjacent to the park? And isn't there a very exclusive golf course resort community further downstream the Wash that undertook some massive water diversion project around something called "Lake Las Vegas"? What is that all about?

Much of the land was undevelopable because of persistent flooding. And yes, some developers are capitalizing on what the park will become by situating their homes in such a way to be able to charge a parkside premium. The Wash is actually diverted underneath the artificial Lake Las Vegas—which predates the current master plan—which is filled with water pumped up from Lake Mead.

One aspect of the project was halted because of some archeological finds. Is there an active tribal presence in the area and how have they been involved? Given that there are very few wet spots in the desert, the Wash area must have been an incredibly important place for them.

There still is an active tribal presence in the Valley, but it must be admitted that during the decade-long process of working in the Wash they have never taken an active interest therein.

There are two local reservations right in the middle of Las Vegas that review the Parks and Recreation Department's work. And we use the Section 106 Cultural Clearances to satisfy the federal regulatory requirements. But beyond that, none of the tribal groups have stated that this is a place sacred to them.

This is such a good project in demonstrating the complexity of restoration by showing that no single discipline can ever hope to solve all the problems in isolation and that rather a team approach is required. What then are some personal reflections on this process and the formation of the "Wetlands Consortium"?

The first decision that called for a lot of discussion was the formation of the team itself. Design Workshop as the landscape architects were the lead on the master planning effort that included the present members as part of the primary team, as well as an additional eleven subconsultants. Montgomery Watson Harza were the lead on the environmental impact statement and the environmental assessment work. But again, the same group of people were involved so the created synergies were critical to fostering a long-term working relationship with individuals who really came to understand the issues. This became especially important given that there has been a large turnover in staff at the Clark County Parks and Recreation Department and the County Planning Office. In other words, we now have a situation where it is the consulting team who represent the history of the project. What happened here is that as a team we really got caught up in the cause and truly believed in the outcome. As a result, all the long-term consulting team members have invested so much personal time that we have gone far past the professional consulting model in that the Wetlands Park is now a passion for all of us. And we have become friends in the process, something that has lasted even with us now being scattered across the country.

chapter 7

*From Barn Elms reservoirs to the London Wetland Centre**

Background and project development

The Wildfowl and Wetlands Trust (WWT), started in 1946, is one of the oldest conservation organizations in the United Kingdom and the only one dedicated to the conservation of wetlands and their biodiversity. With a motto of "saving wetlands for wildlife and people," WWT's central focus has been to put people at the center of everything in which they become involved. This has resulted in a transformation of the operational mandate of the WWT from when ducks and swans were clearly the central focus. The organization was founded by Sir Peter Scott, son of the Antarctic explorer. The latter had written in his polar journal to his wife instructing her, should he not return from the expedition, "to interest the boy in nature...[as there are]...some schools that teach it." The schooling obviously had the intended desire of Captain Scott since his son Peter went on to establish many national and international environmental organizations (such as the World Wildlife Fund for Nature) while making his living as a wildlife painter. He was the first person to be knighted for services to conservation (in 1973) and has been described by the eminent naturalist, Sir David Attenborough, as being the "patron saint of [British] conservation."

Wetland drainage and habitat loss combined with disturbance and hunting pressure continue to threaten wetland biodiversity in the United Kingdom (UK). Scott's lifelong belief was that the best way to stimulate people's concern for wildlife was to encourage close contact with nature. Today, the WWT is involved in conservation efforts worldwide, with a membership exceeding 100,000. The organization is based on getting people and wildlife together for the benefit of both. This is accomplished by developing and operating nine centers in the United Kingdom where people can be intimately exposed to wildlife, both captive animals and individuals collected from the wild. These centers take the form of a mix of botanic garden, interactive museum, and nature preserve, with

---

* Adapted from a presentation by Doug Hulyer, Kevin Peberdy, and Malcolm Whiteside.

education at the inspirational core. This mission is undertaken through employing a variety of techniques such as interactive media, active learning, and interpretive exhibits, such that WWT is internationally regarded today as being one of the leaders concerned with lifelong learning in wild landscapes.

The WWT has developed extensive expertise over the years in how to integrate wildlife into its reserves. In the 1960s and 1970s, the WWT took post-agricultural landscapes and transformed them into wildlife sanctuaries. The objective in those days was to take people from where they lived (mostly cities) to somewhere else (the countryside) where wildlife occurred. Recently, the challenge has been to find new ways in which to transplant such rewarding experiences generated in the countryside back into the city.

In 1988, a decision was made to add the word "wetlands" to the organization's title to indicate that habitat was as important as waterfowl. This was a revolutionary move that occurred at a time when the very concept of a "wetland" was novel. For example, as I relate in the introduction to my wetland design book, one UK scientist in 1994 wrote: "A negative view of wetlands has persisted to recent times. About 15 years ago, I proposed a research project to a federal agency to study the fate of plant nutrients that move from uplands to adjacent wetlands. The project was funded, but the federal scientist who was my project officer had changed the word wetlands to wastelands in his description of the work. He had never encountered the word wetland before and thought I had misspelled wasteland. This same federal agency now supports research on wetlands. In Great Britain, the term wetland was not even in use as recently as a decade ago." The retooled WWT, with education now firmly at the core of its mission, recognized that its established network of centers did not address issues of where most of the populace lived. Taking out a national map, the committee at a planning meeting asked the question of where they would like to develop the next center. Almost uniformly, all put their fingers right over London. Much later, the WWT would capitalize on this concept of urban wildlife by carrying the catchy phrase: "London Life...Just Got Wilder" on the cover of their brochures about the London Wetland Centre as well as on local buses (Figure 7.1).

Recognizing that London is a global city with its own problems and its own attractions, the committee also realized that it is, of course, an important center for people. With a new policy goal to reengage people with wildlife, London became the obvious choice for a new center particularly given that that was where many of the nationally important decision makers lived, those who with a stroke of the pen on a decision in Parliament can make a massive difference to the lives of both humans and wildlife. London also made sense as the desired location because of the presence of many urban wildlife programs in schools as well as many people generally interested in nature fostered through home gardening.

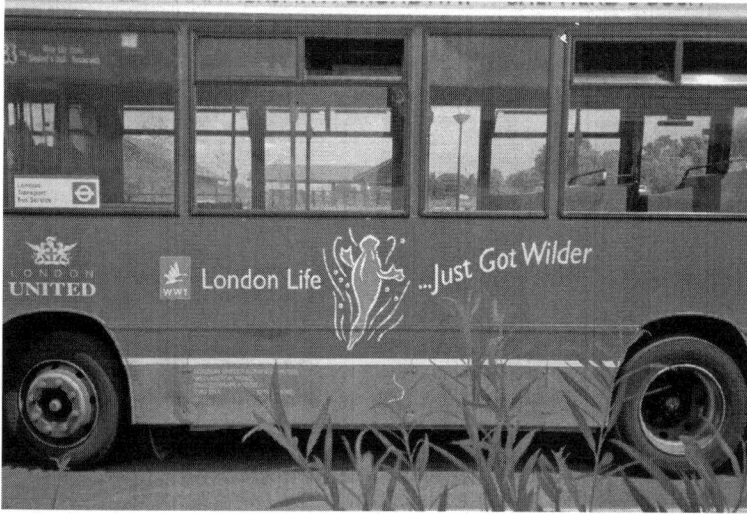

Figure 7.1 Advertisement on side of London bus about the London Wetland Centre.

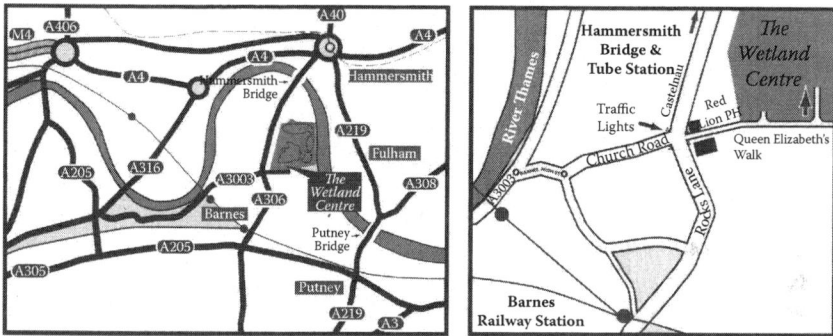

Figure 7.2 Location of the London Wetland Centre.

The WWT first looked at the massive redevelopment opportunities in the Docklands area in East London but could find no obvious place for a natural wildlife center in a location where both the public and government wanted hardscapes to which to moor boats. Meanwhile, a visionary individual working for Thames Water, the public company responsible for water supply and sewerage along the length of the river, had heard about the search for a wildfowl center and suggested the company-owned, derelict property of Barn Elms Reservoirs.

The Barn Elms Reservoirs, located in West London in the suburb of Barnes (Figure 7.2), was comprised of four brick and concrete-lined reservoirs built in the late nineteenth century with hand labor (Figure 7.3). The great pits were dug 9 m deep to supply water to the west of London.

Figure 7.3 Aerial photograph of Barn Elms Reservoirs prior to project development.

North filtration beds and waterworks facilities were situated. The Thames Lee Tunnel carried freshwater from the River Thames west of London to reservoirs in north London via the Barnes site. Over time, however, Thames Water recognized that because the aboveground reservoirs could never supply the needs of the growing city, a massive program called the Thames Water Ring Main—another enormous tunnel—was built to supply water to London. As a result, the aboveground reservoirs became largely redundant as a source of potable water.

Before the reservoirs existed on the site, visitors from downtown London such as Samuel Pepys in the mid-seventeenth century would take boats up the Thames to Barnes where the latter "walked the length of the Elmes, and with great pleasure saw gallant ladies and people come with their bottles and baskets and forms under the trees by the waterside which was mighty pleasant." (Figure 7.4). Though London continued to grow and spread westwards upstream, the large bend in the River at Barnes continued to remain an open strolling and agrarian space until selected as the location for construction of the drinking water reservoirs in 1897.

Because of nineteenth-century safety concerns the reservoirs were surrounded by enormous surrounding berms, one of which was even higher than the subsequently built adjacent housing (Figure 7.5). Because of the risks involved in perching water so high, the abandoned reservoirs required constant maintenance. The other important observation was that ducks liked to use the site for roosting at densities that were nationally

Figure 7.4 The Thames bankside at Barnes in the seventeenth century.

significant. With this in mind, the government, supported by WWT, had the location designated as a Site of Special Scientific Interest (SSSI) for conservation. The result was that Thames Water couldn't substantially change the open-water character of the site. Added to the continual maintenance problems, the site had become a major commercial problem for Thames Water to the extent that they were looking for an innovative solution to get out of the quandary. As a result, in 1988 the site was offered to the WWT, which assembled a project development and implementation team composed of hydrological and civil engineers, natural and physical scientists, architects and landscape architects, and educators and interpreters. The main development principles were to integrate a diversity of wetland wildlife habitat types with visitors through the innovative design of site amenities and exhibitory to educate and enthuse visitors about wetland conservation. The reserve area would allow city dwellers to enjoy the natural world, whereas the zoo-like exhibits would enable people to enjoy close contact with some of the world's most fascinating and threatened waterfowl.

The site is located right beneath a major flypath into Heathrow airport and thus was a very well-known landmark for all those who were up in the air. Down on the ground, however, most people didn't even know the site existed. The exceptions were the urban naturalists and children who would sneak in to do some birdwatching. Situated near Hammersmith Bridge where the famous Oxford-Cambridge boat race happens each year, and located within only a few kilometers from the center of London along the Thames Valley bicycle path (Figure 7.6) between Kew Gardens and Kensington and the world famous Natural History Museum, the site was deemed by the WWT to be perfectly placed for the creation of a

Figure 7. 5 View of surrounding berm and abutting neighborhood homes.

wildlife attraction. Sir Peter Scott, representing the WWT, therefore gladly accepted the offer and asked for £15 million to develop the site. Because Thames Water did not have that sort of money, a new idea was born for an "integrated scheme" in which three partners would come together to develop this particular complex.

Figure 7.6 Thames bicycle path abutting the London Wetland Centre.

Because of their wildlife value, the reservoirs had to be retained but could be transformed into new habitats. Because the northern end of the site was found to have a relatively low wildlife value, it was planned to be used for construction of a new housing development. The plan therefore was to have a 30-acre "enabling development" integrated with a 100-acre biodiversity-education-driven center (Figure 7.7). Even for the housing development, the goal was to try to integrate biodiversity features so as to maintain the overall site SSSI status. This was a challenge given that the location for the intended houses contained a pond which supported birds and a rare species of newt as well as old buildings that were being used as roosts and nursery sites for bats.

A detailed site inventory was made to assess the landscape and visual opportunities and constraints as a planning aid. North and west of the reservoirs, water drained toward the water bodies, whereas south of the reservoirs, the surrounding berm deflected drainage toward a small tributary of the Thames (Figure 7.8). Vegetation was restricted to backyard gardens, a group of allotments north of the reservoirs, playgrounds south of the reservoirs, and a forested riparian fringe along the Thames. The area of the reservoirs was assessed as having value for wintering waterfowl and thus designated as a SSSI site (Figure 7.9). Because the Barnes area already suffered from traffic congestion, special attention was paid to mapping and analyzing circulation patterns as a prelude to traffic reconfiguration planning (Figure 7.10). Visual analysis identified regions

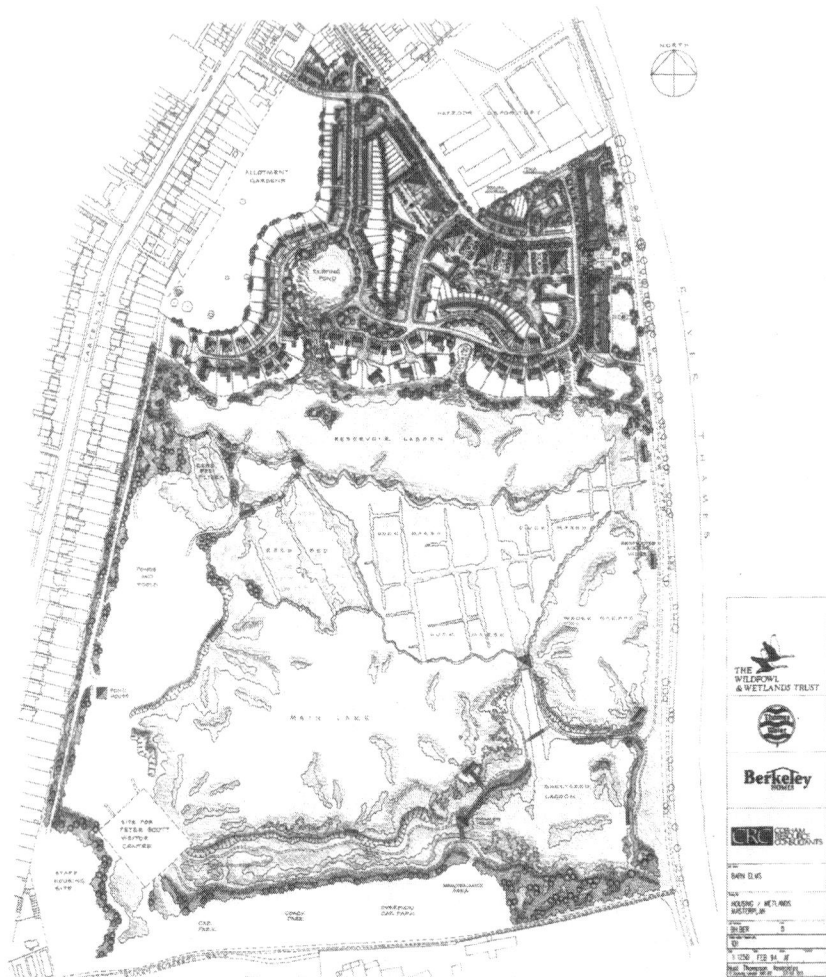

Figure 7.7 Early developmental concepts for the London Wetland Centre.

of urban clutter and open naturalness, as well as sight views to structures or prominent landmarks of scenic interest (Figure 7.11). Sites needing to be screened from view to preserve the feeling of naturalness of the planned nature center could thus be identified.

Permission for planning was sought in the early 1990s and soon hit its first obstacle in that the site was designated as "metropolitan open land," meaning that its open aspect must be maintained unchanged in character. However, an open-minded individual with the local planning authority in the borough of Richmond worked with Thames Water and the WWT to steer the project through a very tortuous planning process based on closely

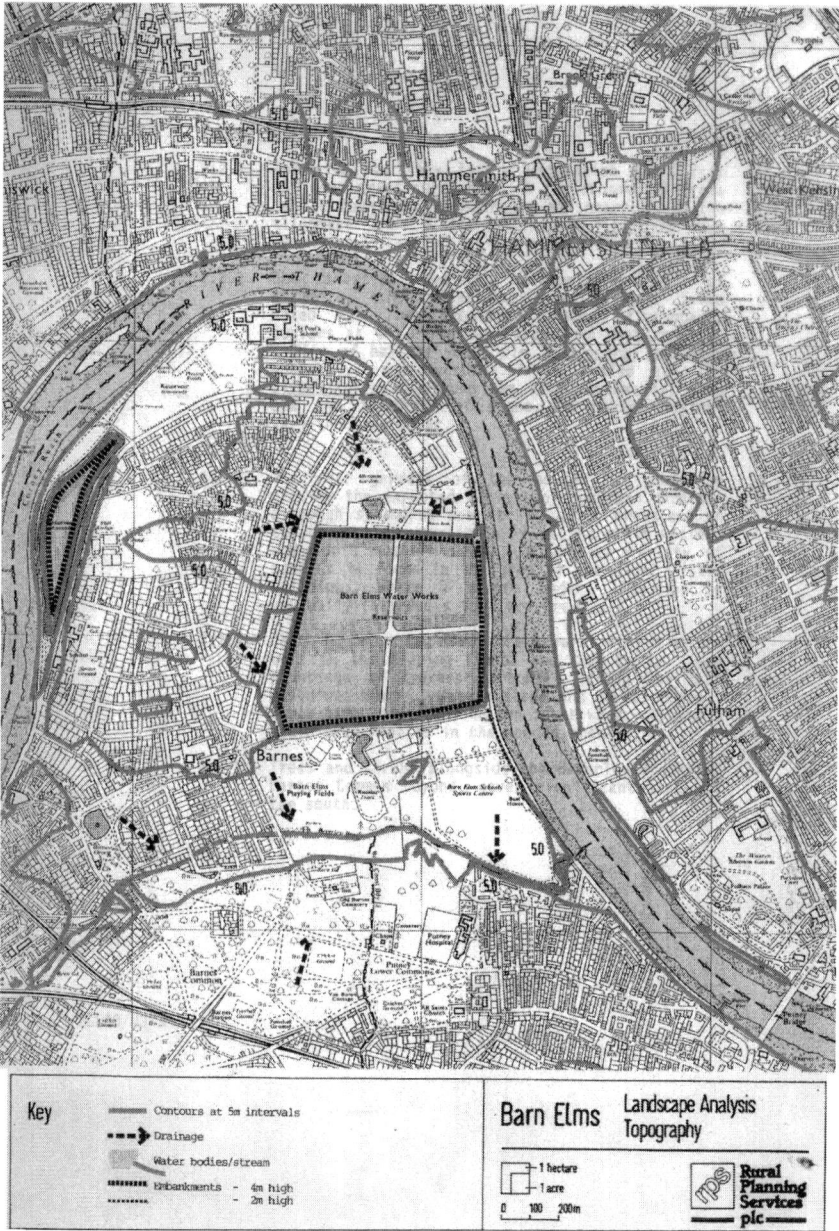

Figure 7.8 Inventory of existing site conditions: landscape analysis of topography.

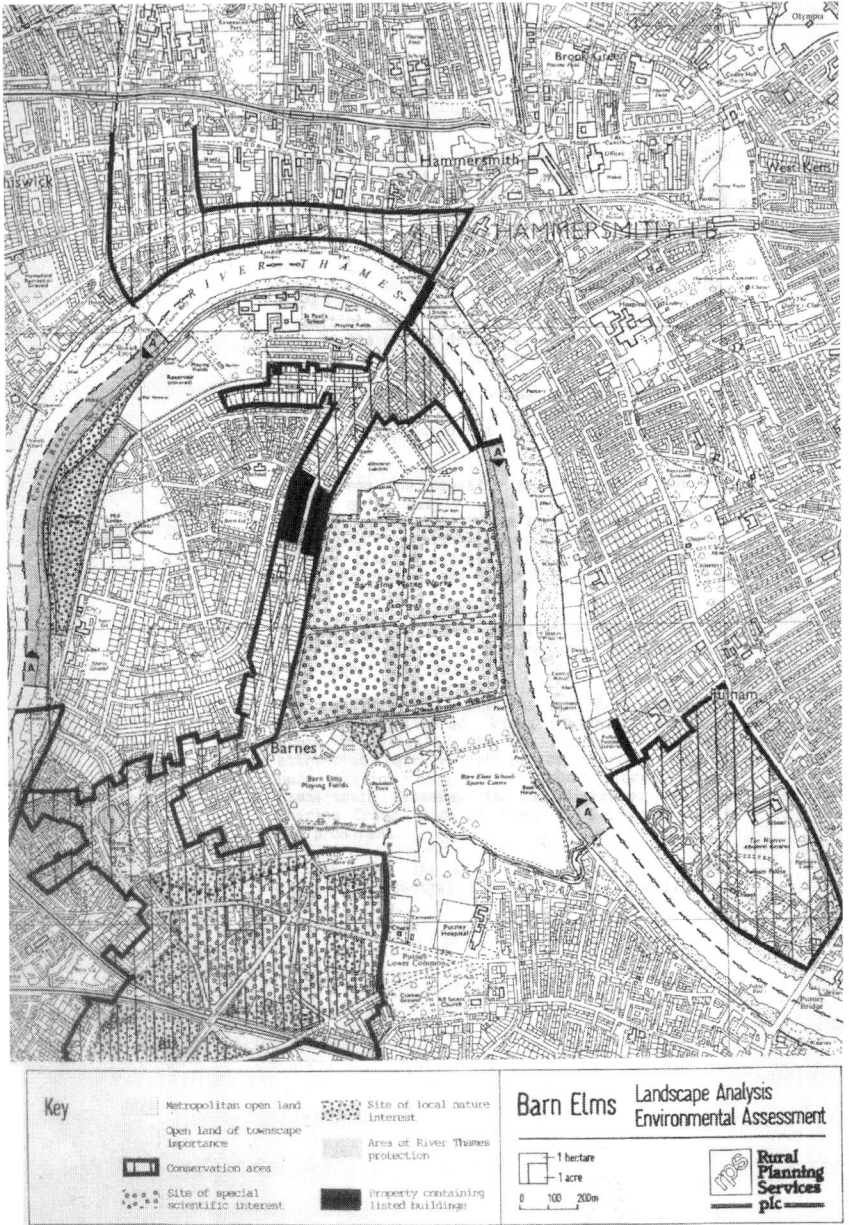

Figure 7.9 Inventory of existing site conditions: landscape analysis of environmental assessment.

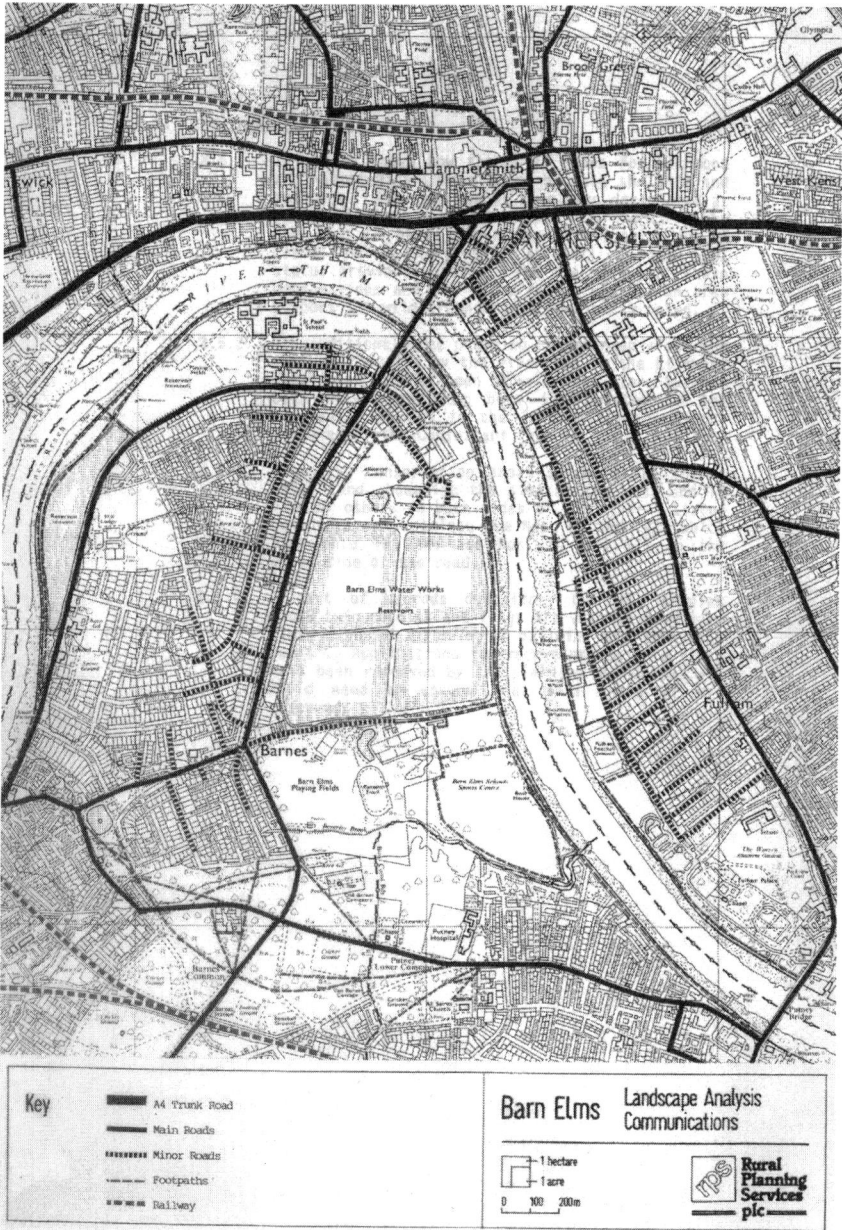

Figure 7.10 Inventory of existing site conditions: landscape analysis of traffic flows.

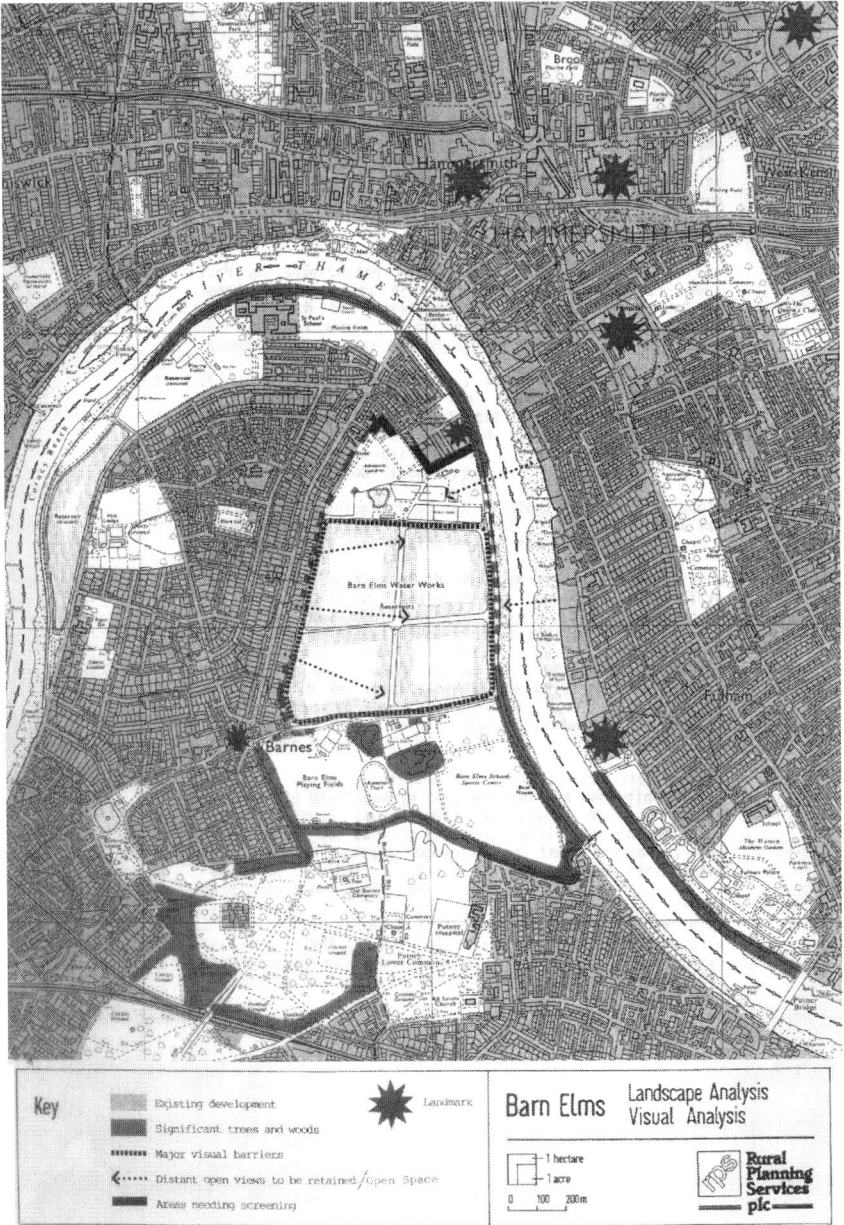

Key

	Existing development
	Significant trees and woods
	Major visual barriers
	Distant open views to be retained/open space
	Areas needing screening

Landmark

Barn Elms Landscape Analysis
Visual Analysis

1 hectare
1 acre

0 100 200m

rps **Rural Planning Services plc**

Figure 7.11 Inventory of existing site conditions: landscape visual analysis.

engaging the local community of southwest London. This was a delicate task from the start for the reason that southeastern England in general and southwest London in particular is one of the most expensive places in which to buy a house in all of Europe. This meant that the abutting community is composed of politicians, media celebrities, and very influential and astute business people who did not want *any* development in their backyards. The WWT therefore adopted the same community participation process that it had successfully used overseas in southeast Asia and other developing locations when dealing with indigenous peoples: talk to them, engage them, and bring them into the process right from the start. The result was partially successful such that the community became supportive of the WWT's wildlife park ideal although they were still against the housing. The challenge then became to persuade the community that the only way to pull off developing an ecological education center was to include the housing component in the plan.

When site development permission was finally granted in 1991, it coincided with a major slump in the housing market throughout all of Britain. This meant that it became a challenge to find a brave developer to take on such a risk. The search took 3 years, until Berkeley Homes signed onto the plan. With a company slogan of "quality to appreciate," this housing developer built luxury homes featuring gold taps, Italian tiling, and places into which to plug your laptop computer in every room. Having experience of building only one or several house groups, this was the first large-scale development undertaken by Berkeley Homes, which has since gone on to become one of the biggest real estate development companies in the United Kingdom.

In the end, the project became a perfect demonstration of the three-part triangle of sustainable development: social, economic, and environmental—in other words, a win-win-win situation. The site has a two-hundred-year lease, much to the chagrin of other, less imaginative NGOs who had previously been offered the location by Thames Water but had turned it down because of the perception of irreconcilable development problems.

Permission for developing the wildlife reserve parcel was obtained in 1991, but it would take another 3 years to both find Berkeley Homes and then to receive the green light for developing the housing parcel. In a strategy of audience management, one "straw man" option that was offered up was to build a series of U-shaped blocks with one massive tower, which of course everyone in the community hated and focused on, thereby letting the more realistic proposal to "sneak" through. The original proposal for 440 units was reduced to 340 and more attention paid toward blending the landscape of the enabling development into that of the Wetland Centre (Figure 7.12).

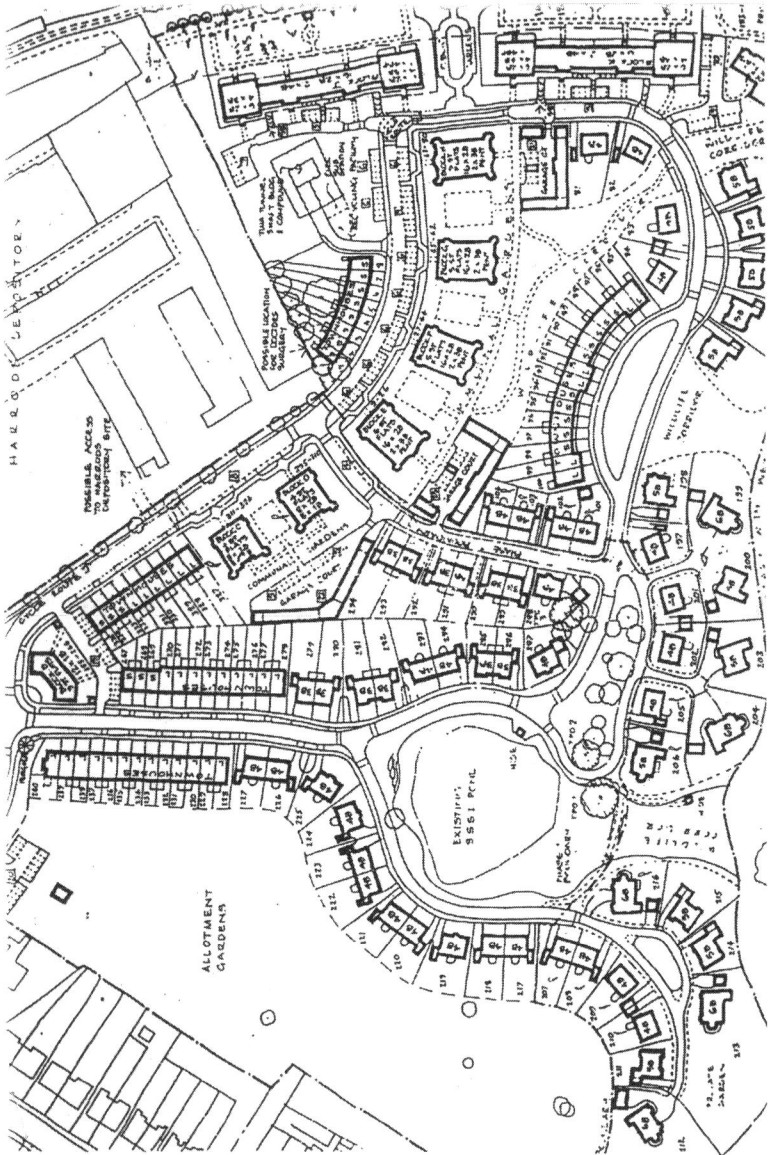

Figure 7.12 Early plans of Berkeley Homes development.

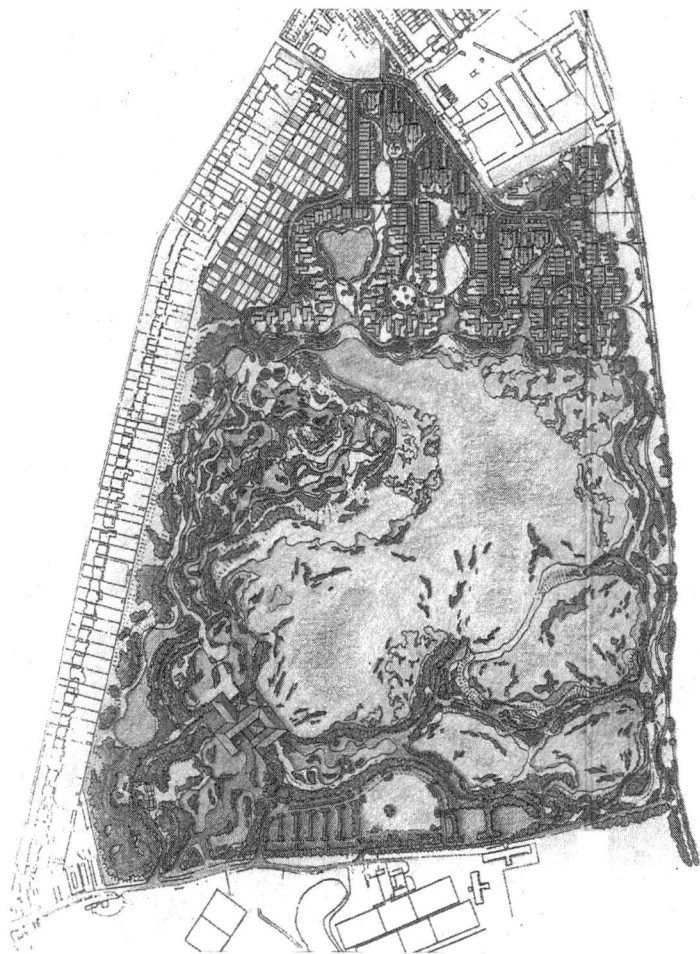

BARN ELMS WILDFOWL & WETLANDS TRUST CENTRE

Figure 7.13 Original plan for the London Wetland Centre showing single central water body.

The constructed wetland in the original plan was to have one large water body and a few much smaller ones (Figs. 7.13, 7.14). Later (discussed below) this scheme was converted to become a mosaic of smaller, different habitat types (Figure 7.15). Site design benefited from the contribution of a landscape architect throughout the project's conception and development. To create a resonance between the two parcels of the overall development scheme, the same architectural firm was hired for both the visitor center building and the Berkeley Homes site.

Figure 7.13 (continued)

The concept for wildlife was to create a whole suite of different habitat types, all specifically designed to serve unique purposes: some for birds, some for amphibians, some for invertebrates, etc. Most of the plan focused on establishing a mosaic of different planting habitat typologies which according to some observers entailed a massive revolution in the mindset of the WWT, an organization which had hitherto concerned itself largely with things (in other words, birds) that were situated on top of the water.

The concept for the people part of the equation was to design an attraction for hundreds of thousands of visitors, some of whom of course would be bird watchers (Figure 7.16). In order to accomplish this objective, it was deemed necessary from the very start to establish a high visibility, major visitor attraction rather than just merely an education center. The focus was therefore on including elements that would draw people in and then once there to subversively educate them.

The Peter Scott Visitor Centre (Figure 7.17) was developed through a design charrette to satisfy several objectives. Though the idea was to have a courtyard that was not a folksy environmental center but something that fitted into the feel of London (Figure 7.18), some early criticism was raised about the functionality of the finished product, an issue that was, however, later addressed.

The bulk of the overall site (70 of the 105 acres) is represented by the "Reserve Habitats," which include the Reservoir Lagoon, Reed Beds, Grazing Marsh, Wader Scrape, Main Lake, and Sheltered Lagoon. The much smaller "Wildside" area is a tranquil place of small pools and ponds designed to support native plants and animals. Two other perimeter areas were established as attractions to lure in paying visitors necessary for the continual operation of the Wetland Centre. The "World Wetlands"

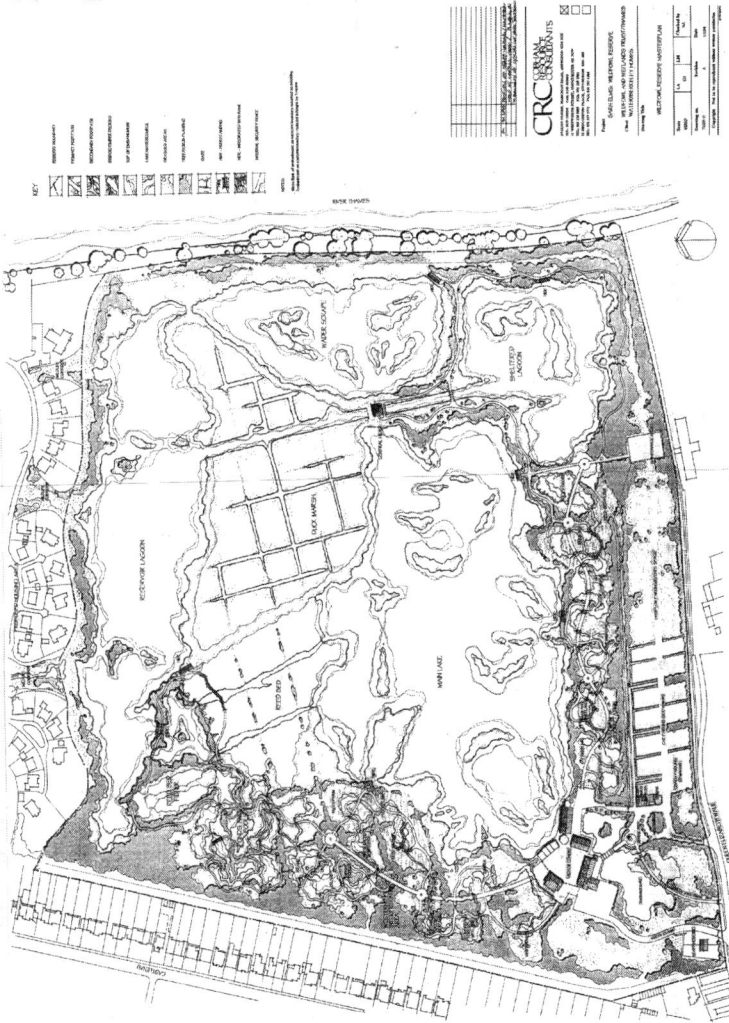

Figure 7.14 Original concept drawing for the London Wetland Centre showing single central water body.

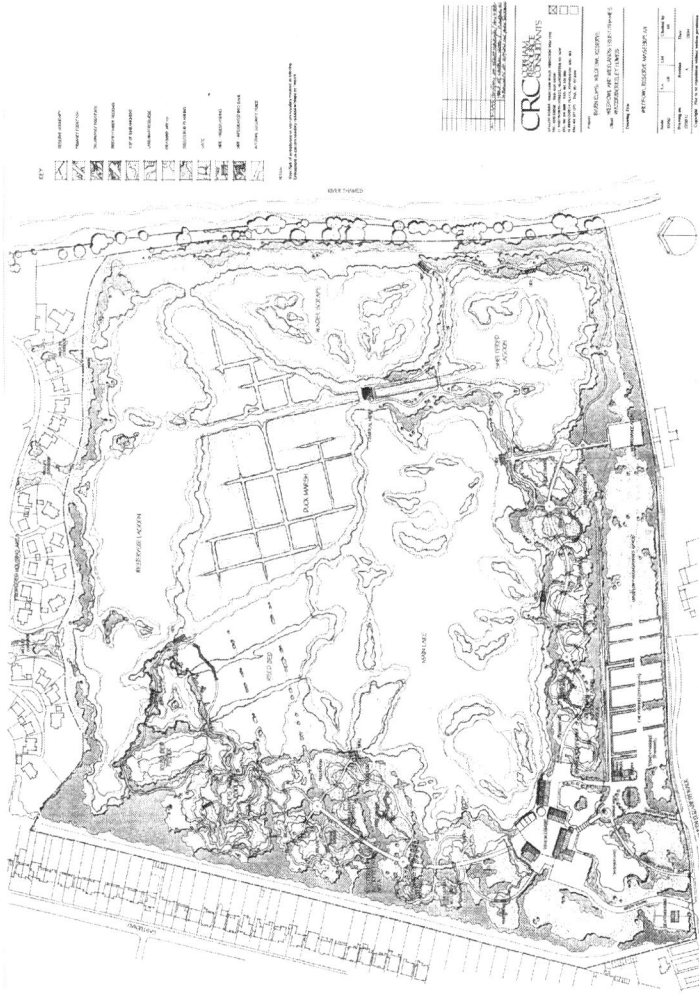

Figure 7.15 Final plan for the London Wetland Centre showing major water bodies of the Reserve Habitats area and the smaller wetlands of the two major outdoor exhibits: Waterlife along the southern (bottom) border and the World Wetlands along the western (left side) border.

Figure 7.16 The London Wetlands Centre attracts many ardent birdwatchers from across the United Kingdom.

Figure 7.17 The Peter Scott Visitor Centre.

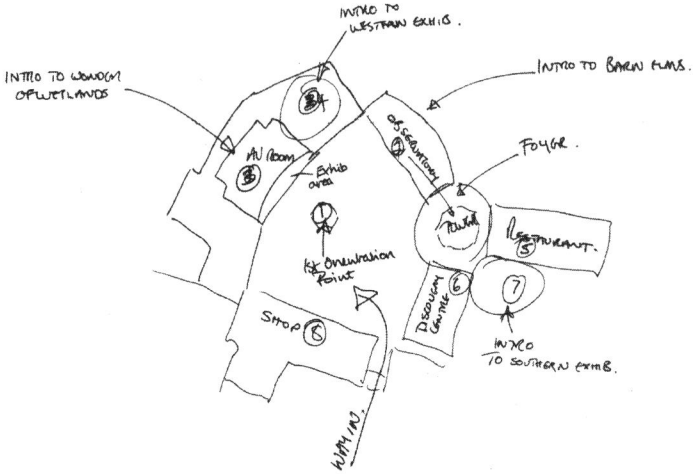

Figure 7.18 Plan and drawing showing central courtyard of Peter Scott Visitor Centre.

exhibit (Figure 7.19) features clipped-winged waterfowl from around the world (Figure 7.20) contained in a series of stylized wetland habitat types representative of their own particular region of origin. Here, within a short distance and separated by gates, visitors can undertake a journey from the arctic tundra, through Scandinavian boreal forest pools, Middle Eastern reed beds, African and South American floodplains, South Asian tropical peat swamp forests, East Asian rice fields, Australian billabongs, Hawaiian lakes, Falkland tidal beaches, and a New Zealand stream containing a highly endangered duck (Figure 7.21). This concept is borrowed from famous American zoos and represents one of the first applications

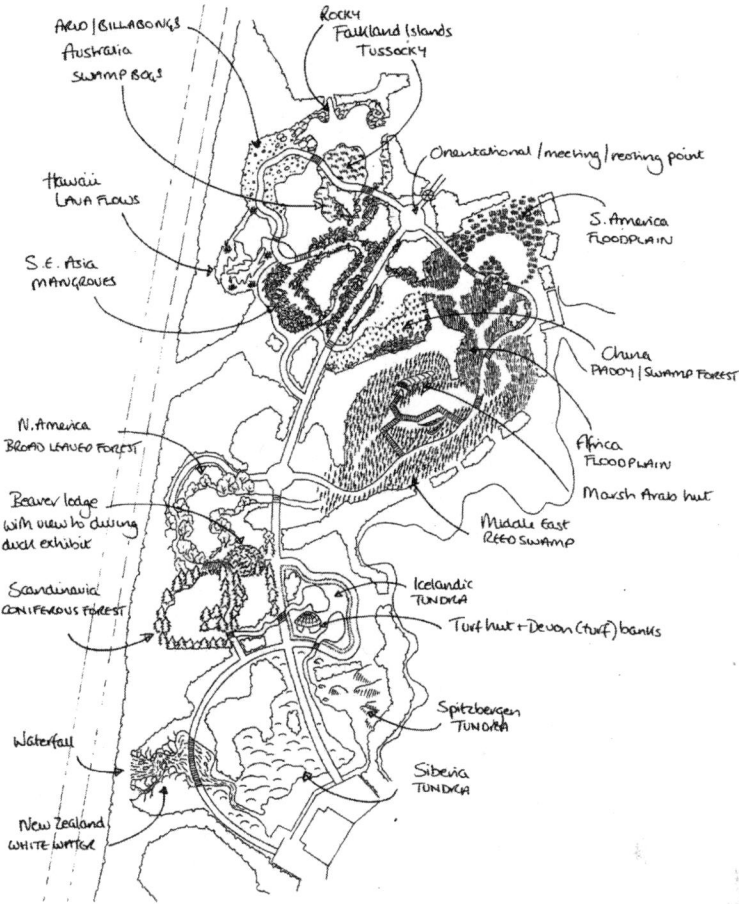

Figure 7.19 Early schematic drawings and map of the World Wetlands exhibit.

of these zoo techniques in the United Kingdom. The "Waterlife" exhibit (Figure 7.22), in which historical and cultural elements such as farming, crafts, and sustainable gardens are located, is designed to inspire people to take a stewardship role in wetland protection.

All homes (Figure 7.23) were sold before completion of construction. This occurred at a time coincident with Hong Kong being returned to China, so Berkeley Homes had its biggest sales drive there. Original prices of £500,000 to £1,800,000 have now doubled in value since the London Wetland Centre opened. The housing development had wildlife corridors built into the landscape plan and the amphibian pond was kept as a sort of traditional village pond. Proximity of these houses overlooking the water bodies (Figure 7.24) enabled an additional waterside premium value to be charged by the real estate developer.

Figure 7.19 (continued)

The housing development ended up generating £11 million to be put back into the project. As this was still short of the total £16 million that was needed, the WWT created an inspired fundraising team who were successful in acquiring the £5 million shortfall in less than 2 years despite many pundits saying that such a task would be impossible given the economic climate of the United Kingdom at that time. Production of a series of brochures and newsletters (Figure 7.25) were important for keeping the momentum.

Sir David Attenborough cut the ribbon on May 25, 2000, at the grand opening, heralding the Wetland Centre as "an ideal model for how the natural world and humanity might exist alongside one another in the centuries to come." And HRH, Charles, the Prince of Wales (Figure 7.26),

Wander through...
World Wetlands
and the Wildside

Figure 7.19 (continued)

citing the importance of the WWT, stated that "practical, sustainable conservation can only be founded on knowledge, and very few doubt that WWT has an unrivalled understanding of the needs of wetland wildlife... It is work that needs the support of all of us who care for wildlife and the wetlands they inhabit."

Figure 7.20 Clipped-winged waterfowl from around the world.

Figure 7.21 Entrance sign, gates between regional exhibits, interpretive signs, and sculptures in World Wetlands exhibit.

Figure 7.21 (continued)

Figure 7.21 (continued)

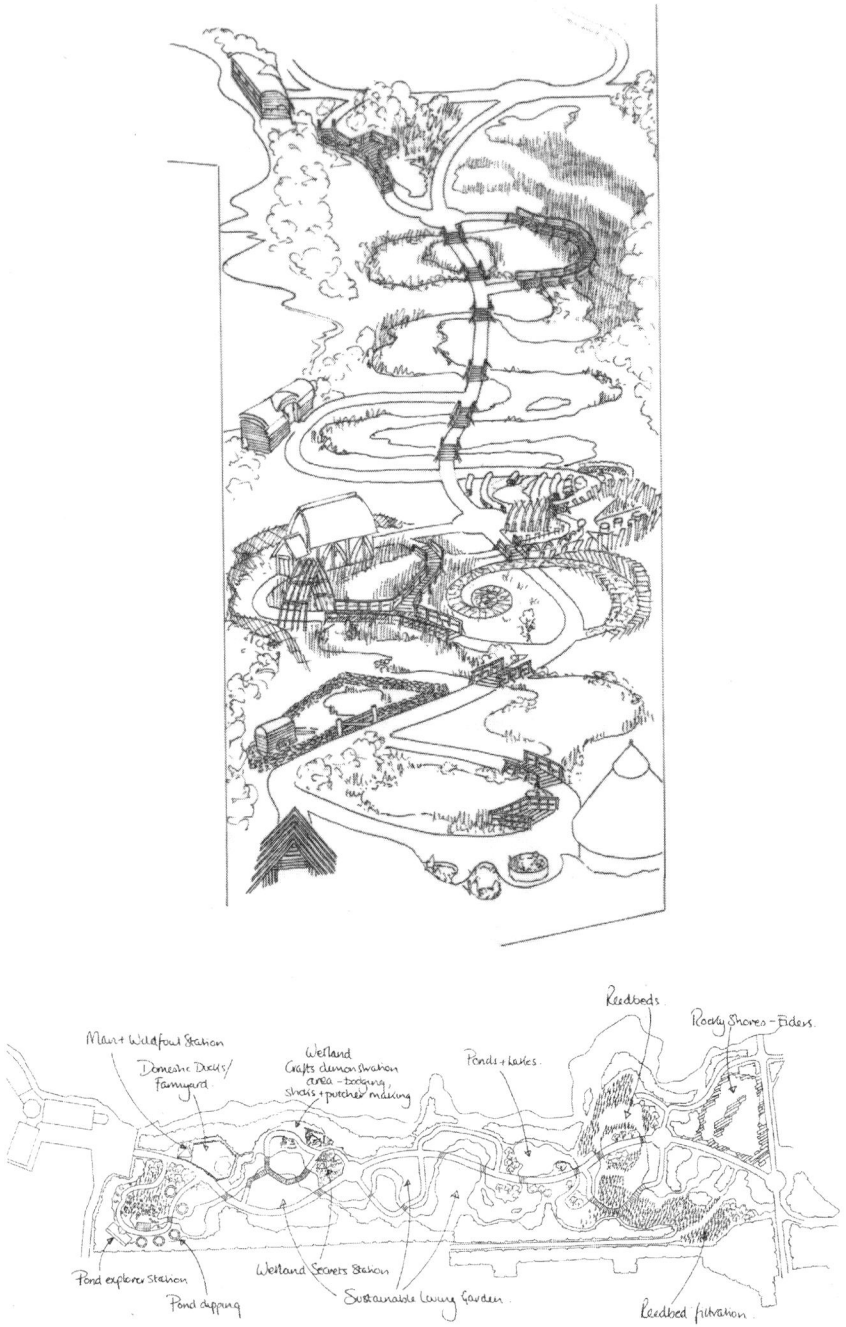

Reedbeds

Rocky Shores – Eiders.

Main Wildfowl Station

Domestic Ducks/
Farmyard.

Wetland
Crafts demonstration
area – bodging
shelts + pitcher making

Ponds + Lakes.

Pond explorer Station

Pond dipping

Wetland Secrets Station

Sustainable Living Garden.

Reedbed filtration.

Figure 7.22 Early schematic drawings and map of the Waterlife exhibit.

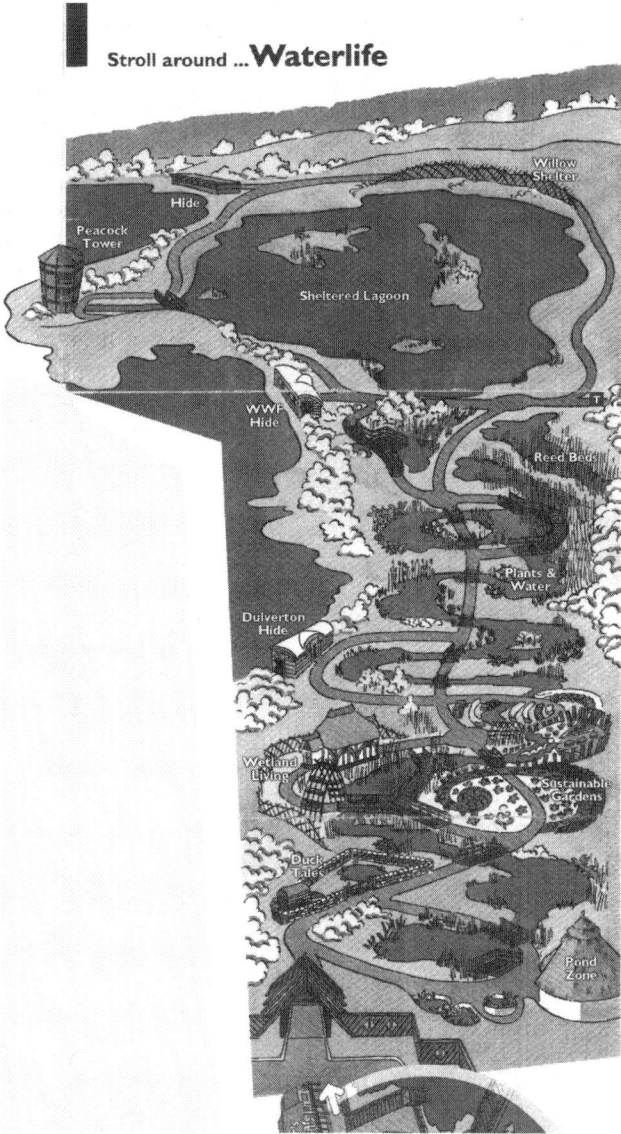

Figure 7.22 (continued)

Figure 7.23 Homes in the "enabling development."

Figure 7.24 Integrated development of the London Wetland Centre with Berkeley Homes.

Figure 7.25 Outreach brochures used for fund raising and notification of project development.

Figure 7.26 His Royal Highness, Charles, the Prince of Wales, visiting the project site during development and escorted by two of the current contributors.

Site design and construction

The Wetland Centre has been called Europe's most exciting and complex natural habitat creation project and is thought to be the largest wetland restoration project in a national capital city anywhere in the world. Several important constraints influenced the actualization of the construction plans. The original reservoir was concrete and brick-lined with a clay core built into each of the banks (and crosspieces) that went down 7 meters to the underlying London boulder clay, thus effectively creating four independent tanks, each half-filled with sand/gravel and about 5 meters of water. All the original soil on the site had been pushed into the embankments at the time of construction. This meant that once the reservoirs were drained, what was left was a huge concrete wall and about 1 meter of silt composed of organic material, heavy metals, and a few WWII bomb shells (some unexploded). A further constraint was that new material could not be brought onto the site to construct the wetlands nor could any old material be removed. Whereas importing construction materials for the new buildings was permitted, everything else had to be won from the site for the wetland construction. The good news was that it turned out that about 70 to 80 percent of the soil needed for final construction could be found in the embankments anyway.

Another problem was the need to take the existing reservoirs out of the Reservoirs Act, which entailed reducing the current height of water relative to the level of the surrounding terrain. What this meant was that the final water levels would have to be at the elevation level of the silt contained in the sidewalls, thereby necessitating the entire site to be dropped in elevation with the new wet-landscape contained within that depressed basin.

A further constraint was the desire to create at least twenty-two different water levels for the various wetlands. The engineering challenge was therefore to convert four simple basins to over twenty new small systems, all linked hydrologically, with the restriction of using only materials that were found onsite (Figure 7.27).

The old filtration beds that had been derelict since the 1960s were to become the site of the new housing development. Here it was important to consider the construction requirements for the housing development such that in the end it was not just volumes of spoil needed for wetlands construction that were important but also the requirement that both projects had to be tightly coordinated in terms of phasing. In other words, certain critical stages had to be reached during the wetland development before all the homes were finished and could be sold. Thus, an important linkage was necessary to ensure that the houses were not developed at the expense of the rest of the site.

The final plan retained the original reservoir walls including the crosswalls, but modified depending on location and the depth of the various planned water bodies. To create the desired number of subbasins (or hydrological units), two main techniques were used. For the larger water bodies, such as the Reservoir Lagoon, Main Lake, Wader Scrape, Grazing Marsh, and Sheltered Lagoon, water was held by further subdividing each of the original clay "boxes." This involved excavating a trench up to 7 meters deep into the underlying natural clay layers and then building a new compacted clay wall up to the required surface water level. For the smaller complexes of water bodies, a second technique was utilized. A clay blanket up to 3 meters below the surface was layered across the entire area with individual perched water bodies separated by short clay cutoff walls keyed into this blanket. Each of these small clay-lined boxes were infilled with suitable spoil and the detailed contours of the water body then sculpted.

The onsite management of the spoil became an extremely difficult logistical undertaking. Not only was the soil excavated from the banks but it had to be categorized and stockpiled separately. This often required double and sometimes triple handing of the soil. Also, because it was not permitted to use any, even temporary, off-site storage areas, this meant

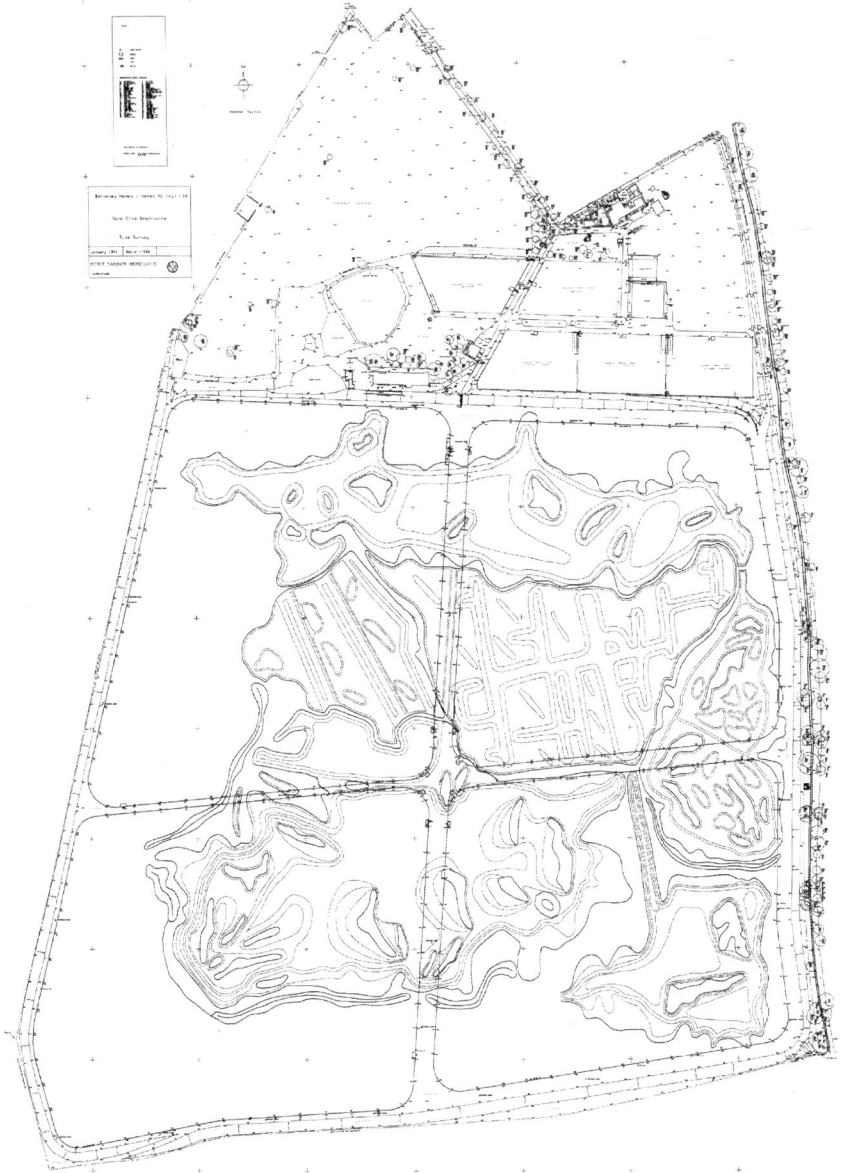

Figure 7.27 Overlay plan of new wetlands to be sculpted from the four reservoir basins.

that all the material had to be constantly shuffled around from spot to spot, out of the way of the most immediate construction location. In total, over 50,000 truckloads of soil were moved. And if all that was not a big enough hurdle to surmount, the other major development complication was the requirement for a water area of at least 20 acres to be present during the entire time in order to maintain the SSSI nature-rating by providing continual habitat for wintering ducks.

The Reserve area of the site is dedicated purely to wildlife. A suite of broad habitat types was created in specific locations. The large open-water Reservoir Lagoon was situated at the northern portion of the site. It is a deep lake containing artificial fish reefs constructed from recycled concrete blocks to attract diving ducks and fish-eating birds, such as herons and cormorants. The Wildside area provides a mosaic of small wetland habitats designed to foster particular organisms such as dragonflies and amphibians. One advantage of lowering the water level over the entire site was that there was now no need to have to rely entirely on pumped water as had been the case for the reservoirs. Instead, it became possible to use gravity feeding. With this in mind, it made sense to locate the lagoon to the north where elevation was highest and then to bleed water from there to the rest of the site (Figures 7.28, 7.29). A central Main Lake was designed not only to replace the original reservoir habitats but to enhance them by increasing habitat diversity. The wildlife potential for birds was increased for purposes of roosting, breeding, and feeding at different water depths.

Great emphasis was placed on developing habitats characterized by different groups of plants. The islands and shores, for example, were planted to provide shelter for summering and nesting waterbirds. At the southeastern corner of the site, the Sheltered Lagoon was created to provide a different open-water environment shaded by trees and having its own particular set of riparian and littoral communities for wintering and molting ducks. Between these two open-water areas are located three sets of seasonal wetlands that have very precise water level control; in other words, at different times they can be flooded or drawn down. These wetlands are comprised of first, Reed Beds in the north-central area, a rare habitat type in the United Kingdom; and second, a recreated Grazing Marsh situated northwest of the big observation tower. This latter was designed to represent the typical historic riparian landscape that would have naturally occurred at this location of the Thames at a point in time when the floodplain was starting to be modified by agriculture through the building of drainage ditches yet still relying on winter inundation to deposit nutrients onto the grazing land. This washland meadow supplies food for sieving dabbling ducks that graze on land such as a goose during the winter, and is drained during summer to provide nesting sites for wading shorebirds. Third, a Wader Scrape was built along the eastern edge of the site. This is a large tract of open mud flats which functions as

Figure 7.28 Aerial view of site showing water movement.

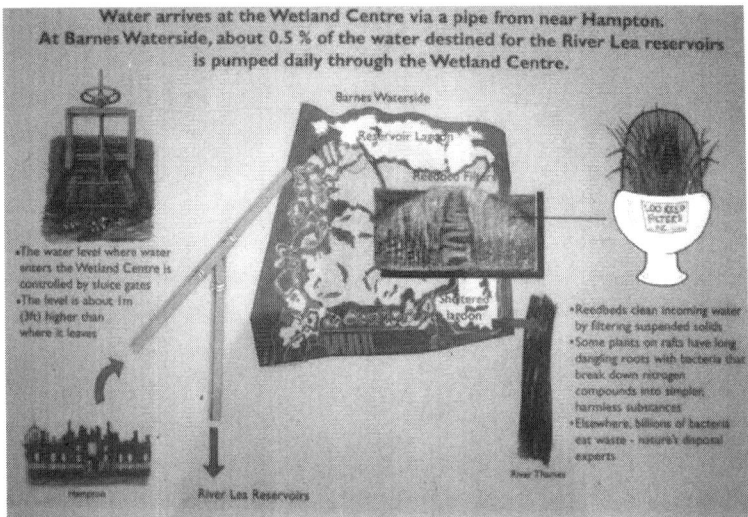

Figure 7.29 Interpretive signage explaining site hydrology.

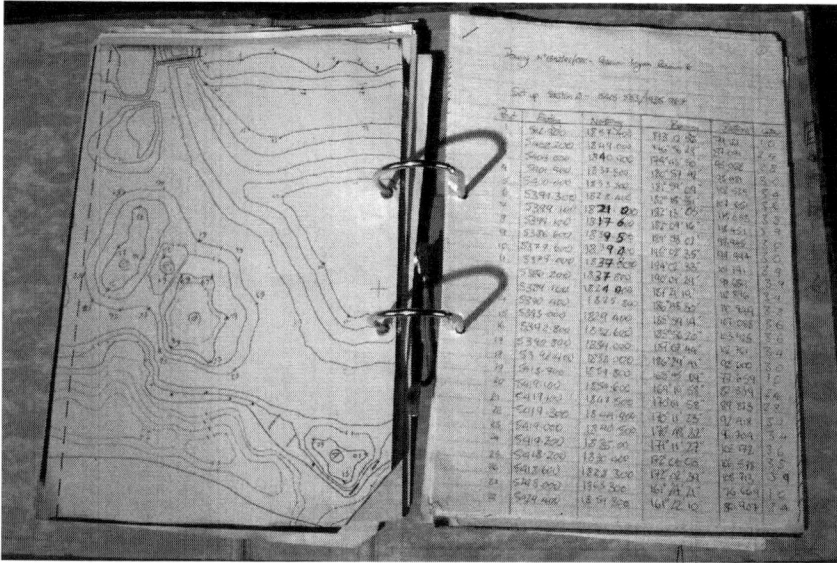

Figure 7.30 Field notes showing detailed elevation recordings during site development of the Wader Scrape.

a rich feeding ground for the wading birds that naturally use the Thames River on the other side of the perimeter berm as a migration corridor each year. Getting the correct and undulating bottom contouring for the Wader Scrape was critical, particularly the very gradual gradient needed to provide a large area of open mud following a small drop in the water level (Figures 7.30, 7.31).

This was a very large-scale construction project which at one time had forty-five 20-ton excavators and huge dumpers working as delicately as possible within a SSSI site. It was necessary to retain the integrity of the outer perimeter berms throughout the entire process to retain their water-holding capacity. Stockpiles of the six different soil types and other material including crushed and blocked concrete can be seen in the bottom left in Figure 7.32, and also the start of Reservoir Lagoon at the top of site. Small shallow bays were sculpted, which would eventually become important focal points for planting within the ditches to create a natural feel to the artificial landscape. The Grazing Marsh was the most difficult habitat to create as never before had such a project really been attempted. This was due to the difficulty in not only trying to create a landform composed of a network of water feeder ditches and ridges and furrows to provide a range of water depths for wintering waterfowl, but also to build a soil profile to allow water to flood the surface and also penetrate the ground, an important requirement for managing such habitats in the

Figure 7.31 Delicate islands in the Wader Scrape.

Figure 7.32 Aerial view of site construction at the stage showing conversion of reservoir basins into the complex patchwork of wetlands.

wild. The variable water depths in the Grazing Marsh are important for attracting a wide diversity of waterfowl; the ridges, for example, which are exposed in the summer after water drawdown provide ideal opportunities for ground nesting birds. Figure 7.32 shows the largely formed Grazing Marsh, the Reservoir Lagoon beginning to fill with water, and a little pool near top that will eventually become the Main Lake. The original clay crosswalls were also used during construction to keep water from flooding into other areas of the site before they were worked upon. In all situations, careful attention to site grading was essential to guarantee project success (Figure 7.33).

Pumps were used for the initial stage of wetland inundation. Ditches were the first to be flooded to check the structural integrity of the entire system by determining how well water was held and what the surface profiling looked like. Wetland creation has to be a two-stage process in terms of obtaining the correct water levels. So first, basic landforming was done for every small pond and lake (Figure 7.34) and these then flooded to a certain level. It then became possible to begin to see what the profile of the shoreline would be, and if it wasn't that which was desired, small machinery or hand digging was used in detailed terraforming for sculpting the shoreline edge. A marginal ring "fence" of deepwater trenches were dug around the Main Lake to prevent the colonization of reeds and other invaders. The islands and shores in the Main Lake were oriented and designed to absorb wind energy. Many islands have deep protected leesides that offer shelter for roosting and feeding diving ducks. As well, the shingle-covered islands in the Wader Scrape were particularly created for the Little-ringed plover.

Hydrologic regulatory structures included simple dropboard sluices with planks slotted in at sides (Figure 7.35) and culverts as a throwback to the old engineering brick of the reservoirs. In situations where extremely precise water level levels were needed, control was maintained through use of tipping-gate weirs (Figure 7.36) where a turn of the wheel allows delicate adjustment of water for absolute millimeter control.

Once the desired water levels over the site were established, it was time for planting. Much of the vegetation used was grown in WWT's own nurseries. All the aquatic plants and trees were natives (Figure 7.37), and wherever possible were obtained from the London area. For example, some of the reed bed seeds were collected from the site itself before construction and grown at WWT's headquarters in Slimbridge and then returned back to the site as small plugs for replanting.

Successfully building a wetland from scratch is a complicated process entailing much more than simply digging a hole in the ground and filling it with water into which potted plants are randomly placed or seeds are sprinkled. One of the first decisions to be faced was whether to deliberately plant the entire site or to let the process of natural colonization do

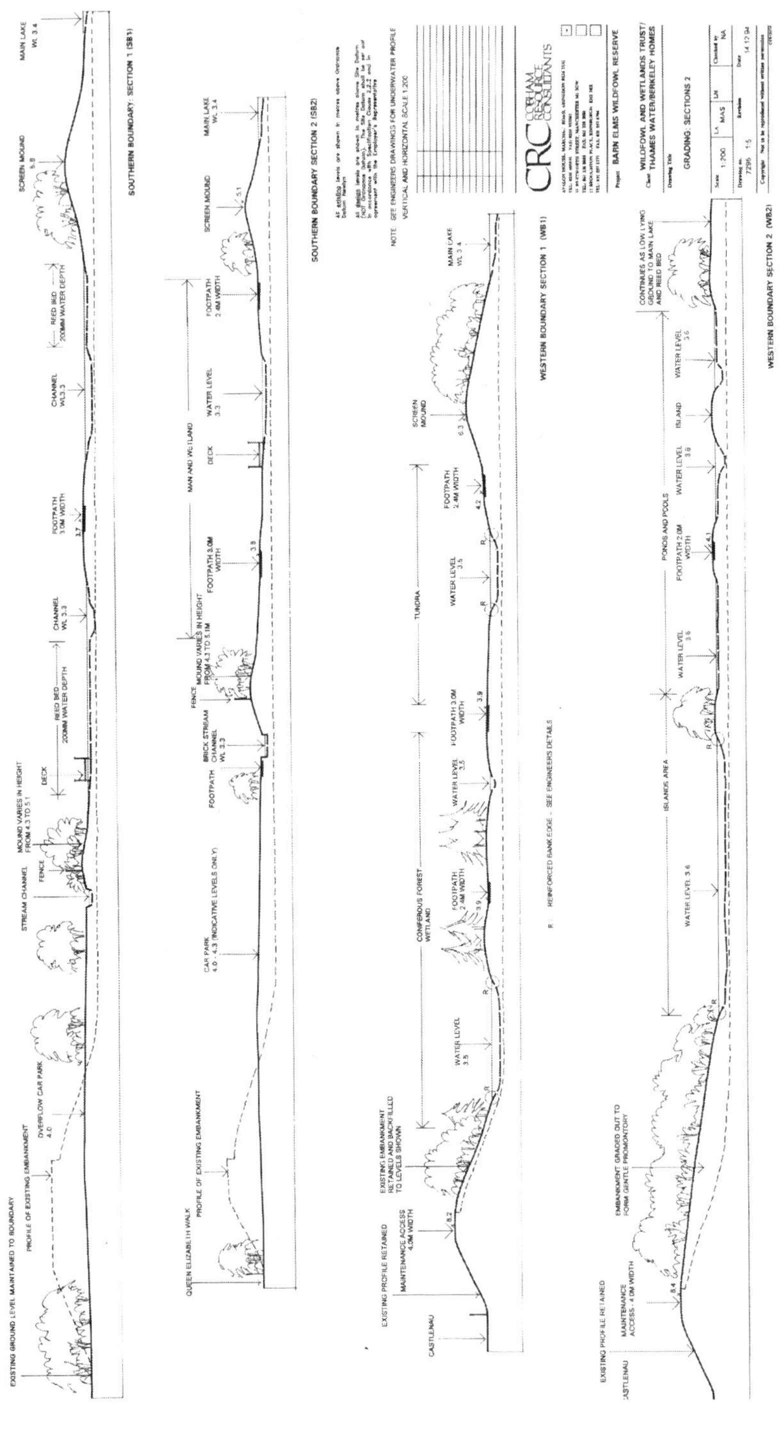

Figure 7.33 Cross-sectional views through water bodies.

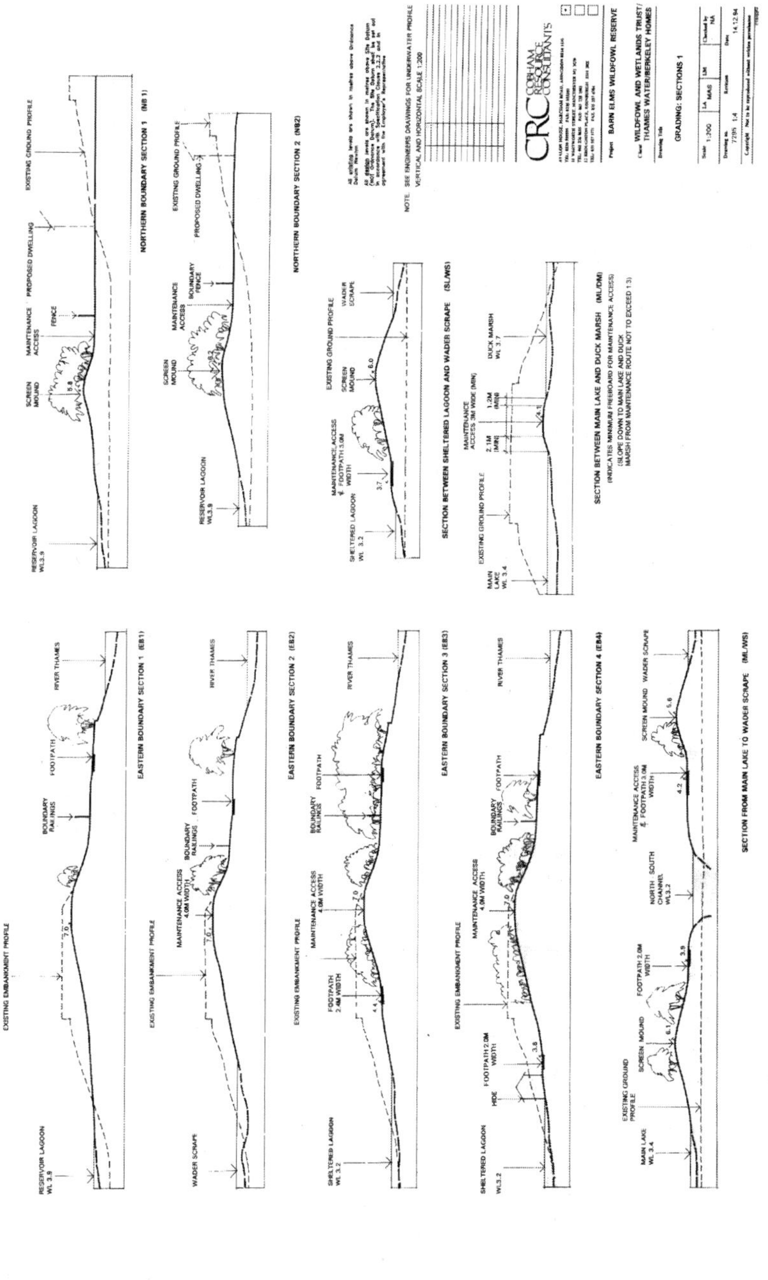

Figure 7.33 (continued)

Figure 7.34 Basic shoreline landforming for the Reservoir Lagoon.

Figure 7.35 A stop-log weir for water control.

most of the work. Given that the site is surrounded by high berms in an urban environment, it was thought that the opportunities for the natural spread of plants would be limited. Also, given the regional abundance of the invasive plant reedmace, other species were predicted to have a difficult time becoming established. As a result, the final decision was made

Figure 7.36 A tipping-gate weir for delicate hydrologic regulation of water levels..

to plant half of the overall site and to leave the rest to colonize naturally. The technique was to plant deliberately or to allow vegetation to naturally colonize small test areas in the site and then to lay down larger areas where final planting would take place with the soils best matched to the known requirements of those particular plants. Given the requirement to work from only the six types of soils occurring onsite, the process was very much like that of an artist's palette where a little of one type of soil would be mixed with another. A considerable amount of preplanning was necessary to work out what the specific vegetation communities would be, where they could be sourced, and then especially to make sure that they arrived to the site at the appropriate time. All incoming plants were screened to remove cultivated varieties and attached alien species.

At the stage when dangers associated with working in a large construction site had passed, volunteers were used for planting, an activity found to be extremely popular amongst a diverse group of interested participants. The original planting scheme used 200,000 plants of 95 different species (Figure 7.38). Aquatic planting took place in single-species blocks in proportion characteristic of natural communities (Figure 7.39). Although the final result is a little contrived in this respect, the London Wetland Centre now represents every type of wetland plant community found in southeastern England.

Because the presence of Canada geese posed a threat to every young plant in the ground, plastic fencing was used to prevent grazing. Although this was recognized to be a somewhat unsustainable approach out of sync with the mission statement the new center was trying to develop, it was

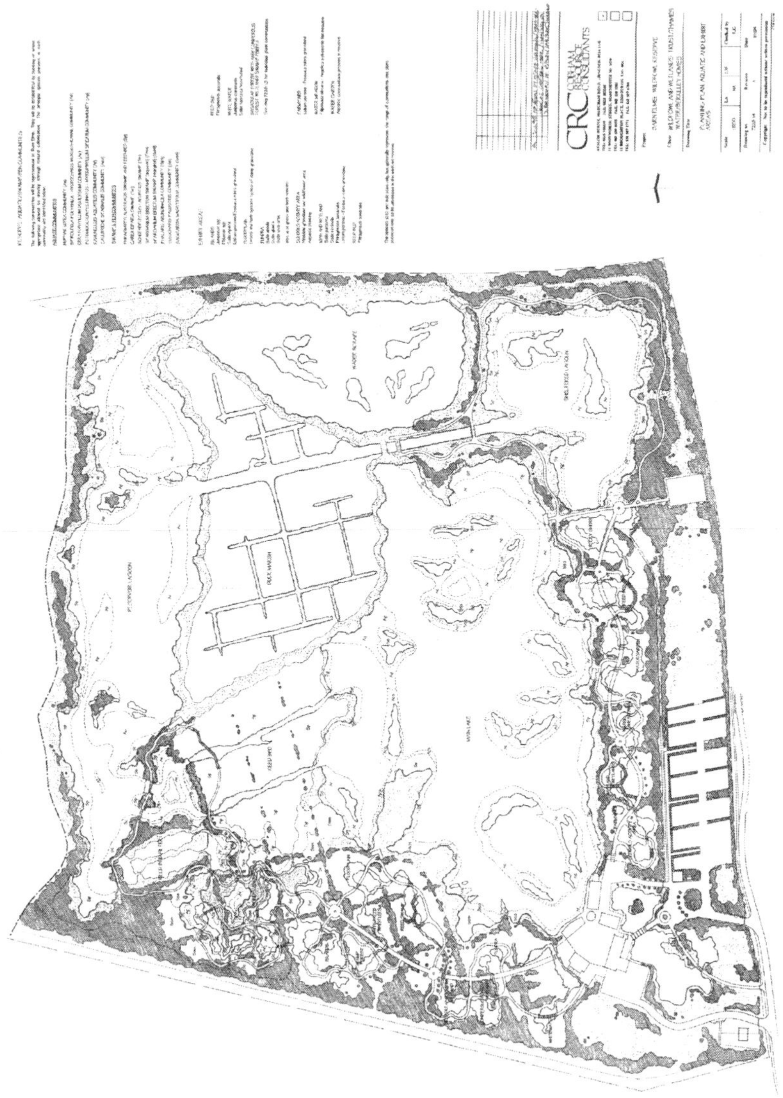

Figure 7.37 Lists of species and their planting plans.

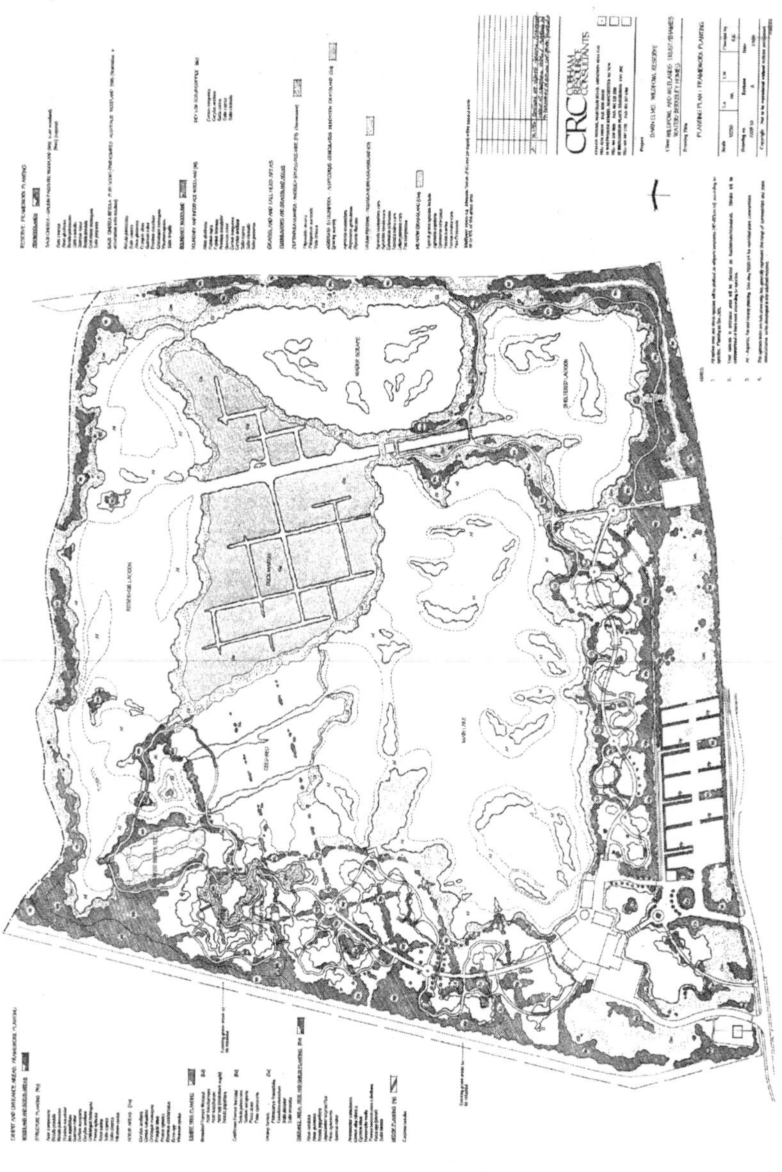

Figure 7.37 (continued)

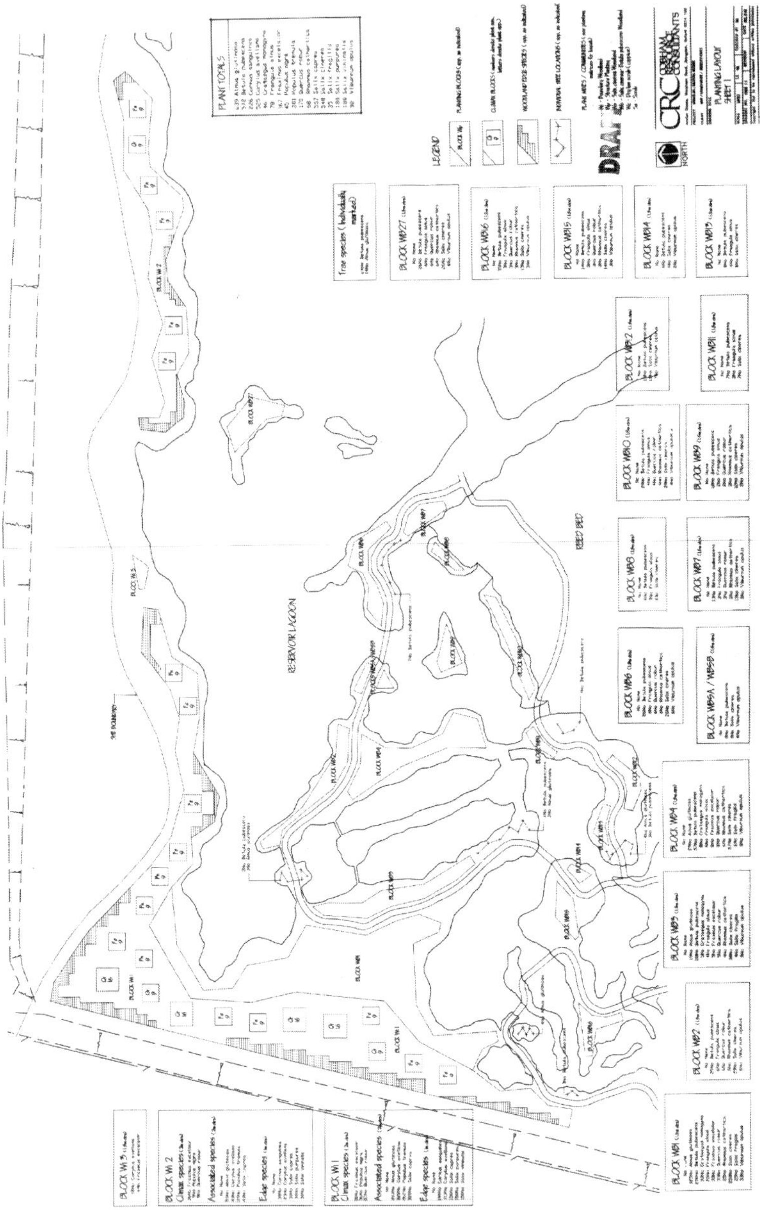

Figure 7.38 Block planting scheme.

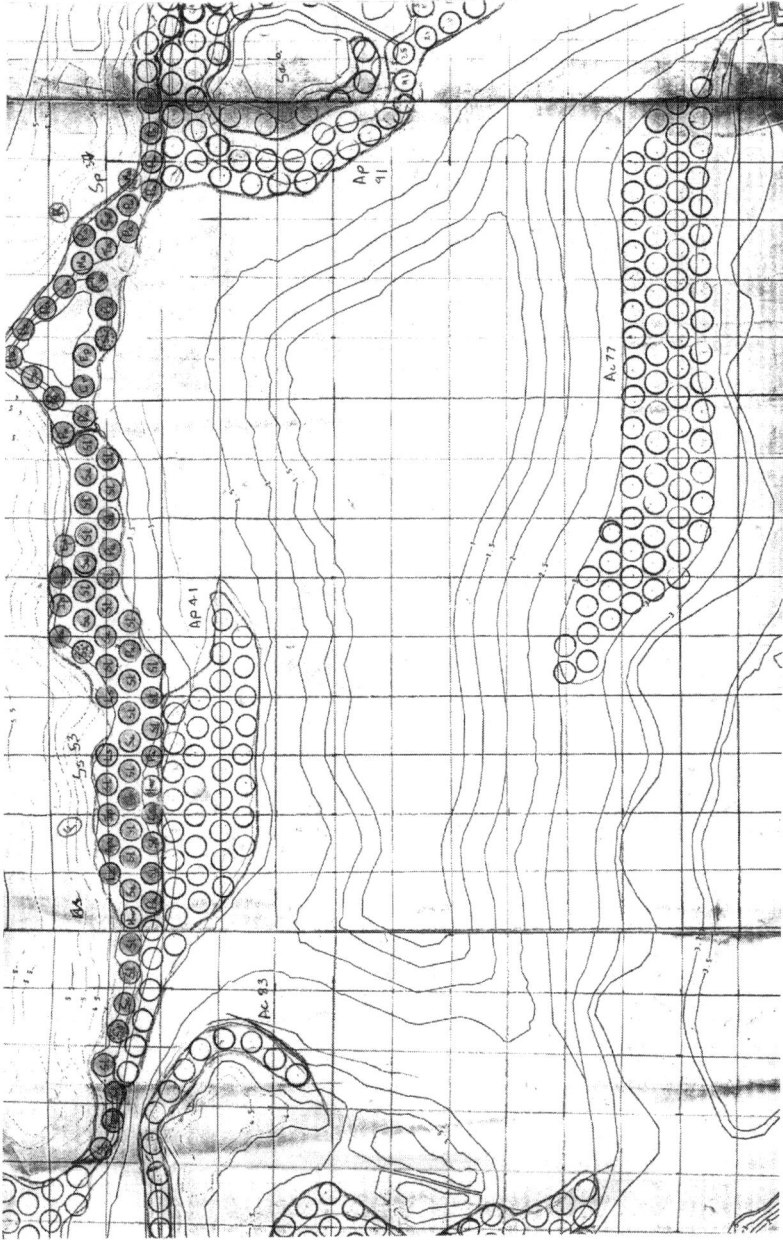

Figure 7.39 Placement plans for wetland and shoreline planting.

Figure 7.40 Aerial view of the London Wetland Centre after site development and prior to opening.

the only viable solution that was light enough to be able to be placed out in the water and inexpensive enough to be employed for the 15 kilometers that was needed. The initial choice of green fencing was soon found to be inadequate as the geese simply jumped over and continued grazing on the young wetland plants. Subsequently, all sorts of other "natural" colors were tried with the same results. Finally, a shipment of bright orange fencing arrived onsite by mistake and this was installed. The result was that, interestingly, no Canada geese touched any plants behind an orange fence. Given that the WWT advises on many wetland creation projects, it is now possible to see this ugly but effective orange fencing going up all over the United Kingdom.

The Grazing Marsh was a particularly difficult habitat to try to establish because the plants needed to be able to tolerate the specific hydrologic regime of winter flooding and spring dampness. To guarantee planting success, workers sought out those regions in the countryside referred to as "washland areas" and collected seeds which were then seeded into the complex surface microtopography of the marsh.

Figure 7.40 is an aerial view showing most of the site after successful development. Many of the features in the Main Lake can be seen, such as the deeper and shallower areas and the islands. Islands and peninsulas were used to create areas of sheltered water in order to establish ideal feeding conditions for diving or dabbling ducks. Each created island was particular with different surface conditions suitable for certain types of wetland birds. For example, some islands were composed of sand or gravel (recycled from onsite) for shingle nesting species like the little ringed-plover and lapwing.

One problem that soon developed was the occurence of blue-green and filamentous green algal blooms during the initial flooding caused by inundating hydric soils with nutrient-rich water from the River Thames. Although summer blooms still do occur, the site has matured and the problem has become much less serious over time. Another problem resulted from the site being gradually handed over from the construction contractors in phases at completely inappropriate times for planting. In one case, the Grazing Marsh became ready with the top soil in place at the completely wrong time of year for grass seeding. As a result, it had to sit in waiting for the entire summer and be constantly managed to keep it free of colonizing species. In some respects this was also the situation for the entire site; in other words, sitting vacant and ready for long periods, covered in blanket weed and orange fencing. The implications of such delays were serious given that the real estate developer was simultaneously trying to encourage potential customers to purchase the £250,000 homes. To counter the nervousness and displeasure, ongoing management was needed and a dedicated and courageous commitment of all parties to circumvent failure. Today the vegetation is thriving.

At one point there were 26 different specialist contractors on the site working on the construction of the various buildings and other architectural features (Figure 7.41). There are 600 meters of boardwalks, 27 bridges, 3.4 kilometers of pathways, and 7 hides, including a two- and a three-story tower. With the boardwalks and bridges (Figure 7.42), the classic mistake was made in terms of leaving their construction until the major groundwork had been finished. The problem that developed concerned underestimating the extent of earthworks required to install these pedestrian features such that their installation actually destroyed significant portions of the surrounding areas. The obvious lesson here is to put boardwalks in first and bring the wetland to them rather than the other way around. Today, the completed boardwalks are tightly integrated into the site, allowing visitor access without disturbing the wildlife (Figure 7.43). Special attention was devoted to the challenge of allowing visitors access to the observation hides without affecting the adjacent wetlands by using a subtle blend of physical and natural screening (Figure 7.44). Additionally, the vegetated roofs on the hides not only serve to camouflage them but also provide feeding areas for small birds and invertebrates.

The viewing areas onsite were designed as a gradation from the massive viewing window at the Visitor Centre Observatory (Figure 7.45) to progressively smaller windows the farther viewers move out into the field, giving the impression of progressing into an increasingly secret place. Then the three-story Peacock Tower hide is reached (Figure 7.46) with its slit windows opening toward the different habitat types radiating

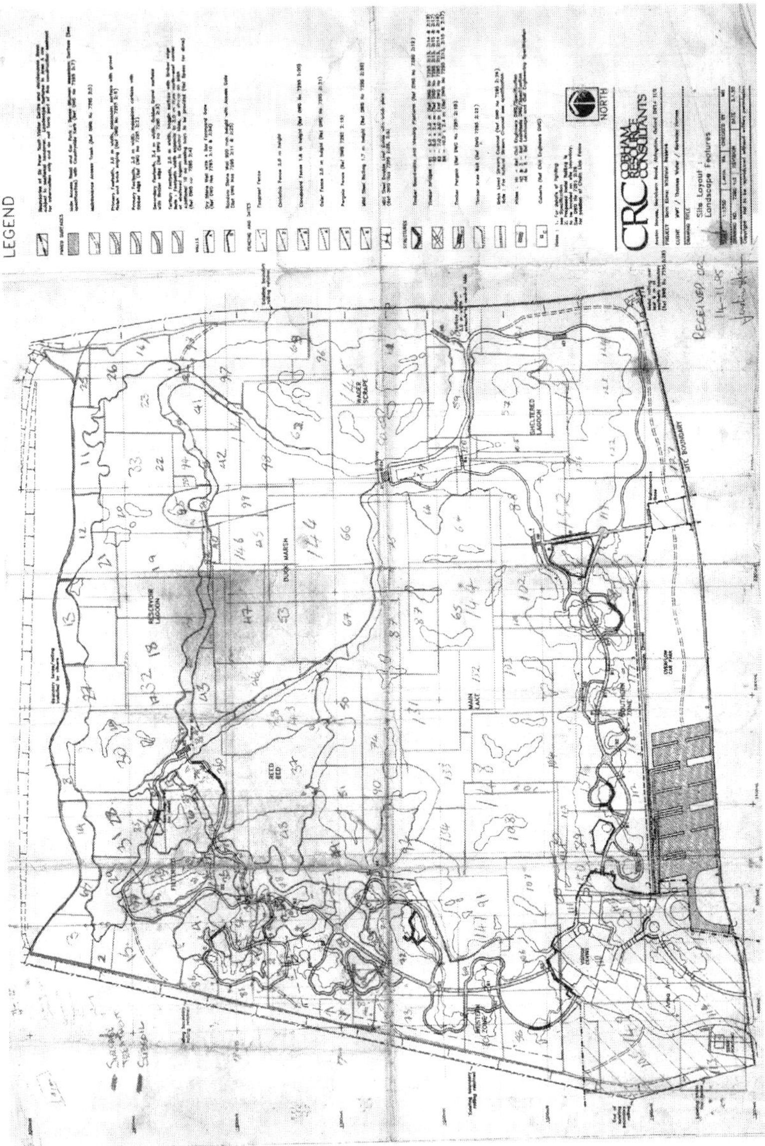

Figure 7.41 Development plans of site architectural features.

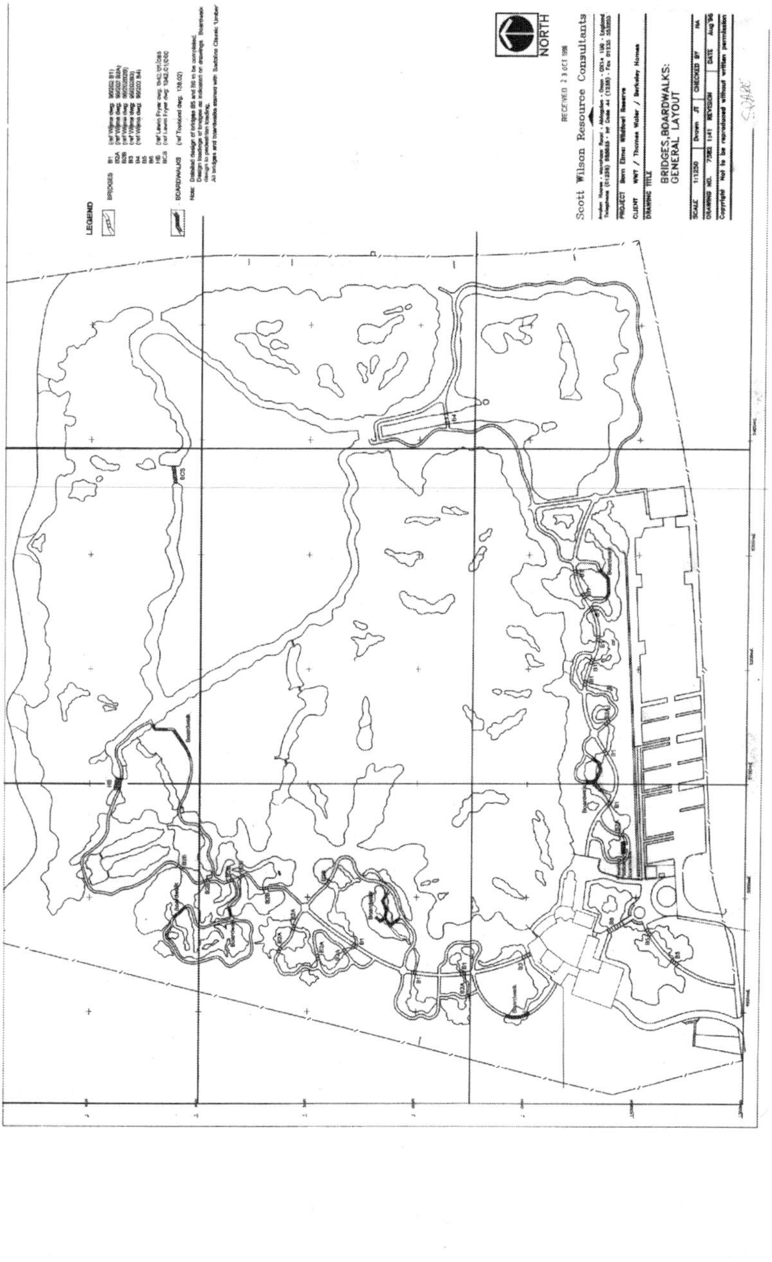

Figure 7.41 (continued)

Figure 7.41 (continued)

Figure 7.42 Representative bridge and boardwalks.

Figure 7.43 Representative boardwalk overlooks.

Figure 7.43 (continued)

Figure 7.44 Representative viewing blinds.

Figure 7.44 (continued)

Figure 7.45 Two-story observation viewing window at the Peter Scott Visitor Centre.

Figure 7.45 (continued)

Figure 7.46 Observation tower.

out from this tower and providing wonderful opportunities for wildlife interpretation and study. This building is the only three-story observation hide in the world that contains an elevator. Such servicing was possible since it was planned from the very start of the project. For example, the entire site is covered with a network of fiberoptic cables (sponsored by a local company) running underneath the footpaths that go out to all the remote cameras (Figure 7.47), which can then be controlled from either the main building or the towers. The cameras focus on secret bird nests at a distance from hides and beam those images not just back to the hide but also back to the Visitor Centre as well.

For the World of Wetlands exhibit, it was necessary to recreate 14 wetland scenes of various locations including a Falkland Island beach complete with dead seal, a Siberian tundra with contoured small hills (pingos) and half-submerged ersatz mammoth tusks, African floodplains with hippo tracks and flamingo nests as interpretive "signs of life"; a New Zealand set of white water rapids, an Australian billabong with lurking (model) crocodiles and platypus burrows, and a southern Iraq marshland.

The concrete that was recycled on site was used for three main purposes: first, large blocks were employed for below-water armoring on peninsulas and islands customarily threatened by undercutting from waves; second, crushed material was used to provide a sub-base for all footpaths; and third, the same material was also used as the foundation for a naturalized overflow car park. For the latter, the desire was to cover the

Figure 7.47 Remote camera mounted in an observation tower.

surface with plants given that it would only periodically be used for parking. A series of experiments were undertaken to determine the most appropriate mixture of crushed concrete and very thin sub-soils found on site. As the resulting mixture produced a change in pH to that characteristic of a natural chalk soil, a limestone grassland was planted. This turned out to be ideal in that such communities are actually dependent on disturbance for their sustenance. As a result, the occasional use of automobiles exposing raw substrate becomes a positive management act in that more skid marks result in more wildflowers that will grow the following year.

The development of the emergent plant communities has been spectacular, most now displaying distinctive seasonal cycles (Figure 7.48) and requiring diligent and ongoing management input (Figure 7.49) that was built into the planning project and budget from the very start. A very important take-home message that cannot be reemphasized enough is that there is no point in trying to tackle a project of this complexity and magnitude without factoring in the ongoing requirement for management and remediation works.

Today, the wide diversity of habitat types has fostered the development of a wide range of wildlife. For example, there are now about 19 species of dragonflies and damselflies visiting the site, of which 14 have been detected breeding. Approximately 170 bird, 350 moth, and 24 butterfly species are recorded annually. Four to 5 species of amphibians and 6 bat species now occupy the site which is recognized as being one of the top two feeding locations for bats in the Greater London region. One of the crowning moments came with the discovery that the reed bed

Figure 7.48 Views of the Grazing Marsh.

area produced 90 pairs of reed warblers when only a single pair had been found there previously. Other locations on the site have been designated as nationally important areas for different species of wintering birds, and more importantly, for certain breeding birds. Over 20 species of birds are now breeding regularly at the Wetland Centre (Figure 7.50), with the site being the most important London breeding location for lapwing and

Figure 7.48 (continued)

Figure 7.49 Ongoing maintenance removal of colonizing plants.

little-ringed plover. Even the WWT, certainly no stranger to the "build it and they [the waterfowl] will come" concept of wetland creation, has been taken aback by the rapid demonstration of success within only a few years of construction. The increase in abundance of dabbling ducks, for example, has been marked, with some species, such as mallards and gadwalls,

Figure 7.49 (continued)

doubling or quadrupling their numbers in the first 4 years of operation. That this has been accomplished in the middle of a city, sprawling over 1,500 square kilometers and home to over 7.5 million human inhabitants and many millions of automobiles, is truly remarkable. Many of these bird numbers are of national importance and have easily led to the site's recognition of a key SSSI in 2001.

Figure 7.50 Resident birds and their habitat at the London Wetland Centre.

Figure 7.50 (continued)

Site operations and education program

The London Wetland Centre is designed to be a place in which visitors, human and waterfowl, can learn or live (Figure 7.51). Until now, and as is the case for all of London's museums, many visitors come from area schools (Figure 7.52). Education should be cross-sectoral and cross-curricular, in

Figure 7.51 Popularity of the London Wetland Centre for education and relaxation.

other words not just about science and nature but also arts and history. The London Wetland Centre is firmly rooted in concepts of environmental education or education for sustainability, which means that it needs to be interactive, lively, and fun (Figure 7.53). For the latter, although the entertainment value is important, there must be a conscious attempt to avoid the "Disney-fication" of nature. The London Wetland Centre is interested

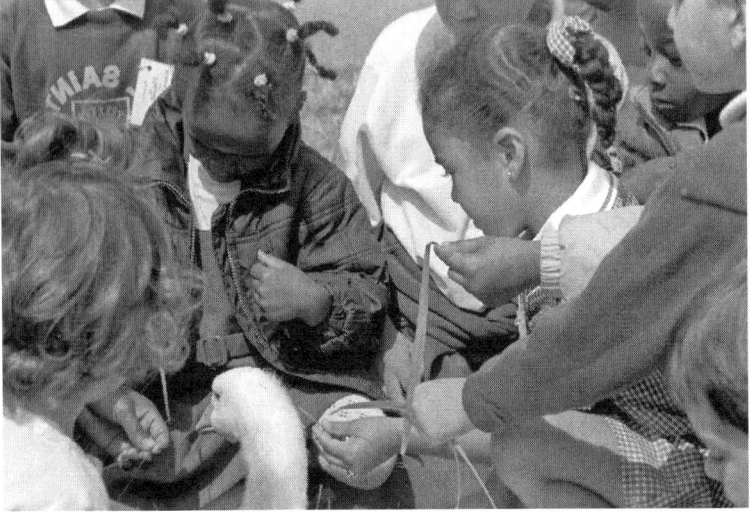

Figure 7.52 Use of the London Wetland Centre as an outdoor classroom.

in both cognitive learning and experiential learning; in other words, what excites people and what can inspire a feeling of "sense of place," as well as "an awareness to action," both important elements for fostering environmental stewardship.

The educational operating agenda of the London Wetland Centre is based on an overall mission triptych of the "five secrets" for each of

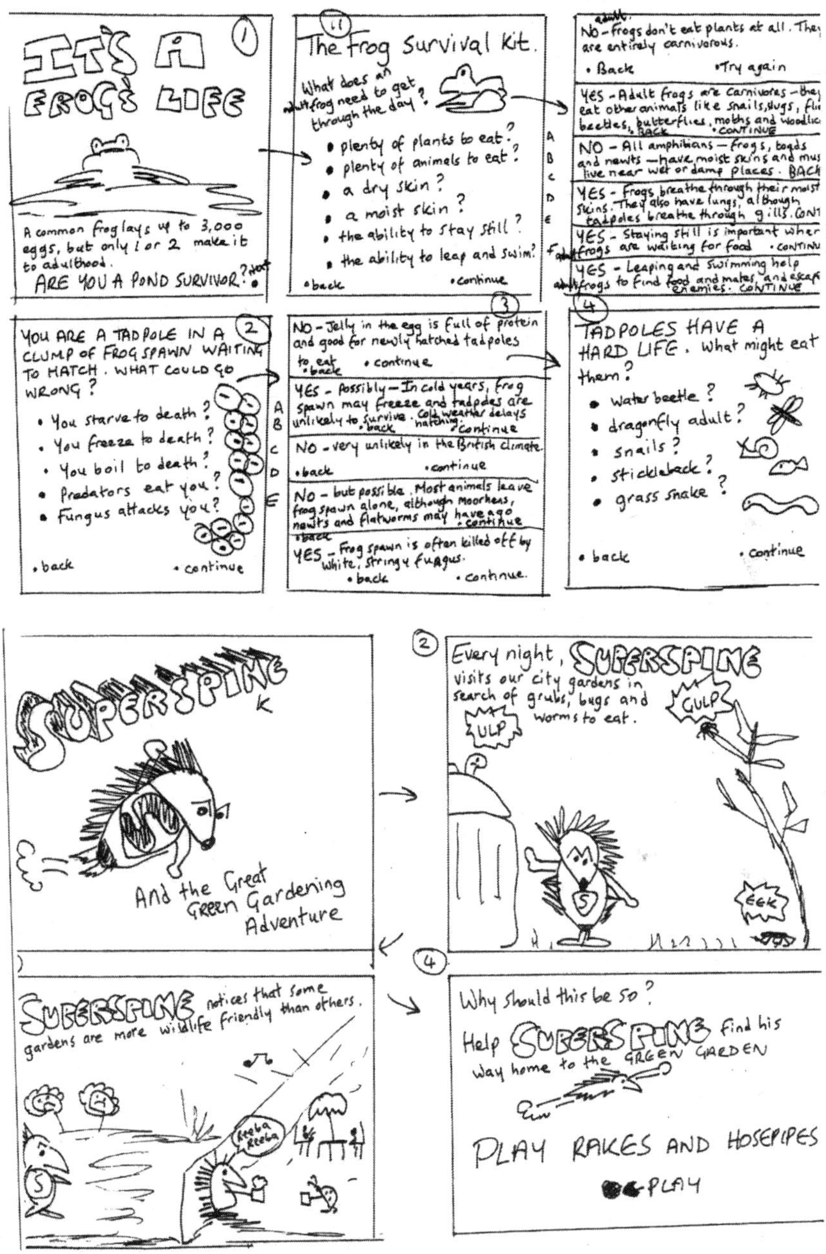

Figure 7.53 Early development of informative and humorous signage for children.

wetland benefits, general ecology, and environmental sustainability. The five secrets of wetland benefits are water storage and flood prevention (in other words, wetlands as landscape sponges), water filtration and purification (i.e., wetlands as landscape kidneys), wetlands as calming refuge for solace and escape from hectic urban lifestyles (i.e., wetlands as refuges for solace), wetlands as production sources for many useful products such as roofing thatch, etc. (i.e., wetlands as nature's larder), and wetlands as centers of biodiversity (i.e., wetlands as oases in a concrete desert). The five secrets of ecology are energy, cycles and material flows, habitat conditions, and the assembly of organisms into biotic communities (for example, Figure 7.54), and the always defining element of dynamic change. And finally, the five, perhaps not-so-secret, "r's" of environmental sustainability are reduce, reuse, recycle—often repeated as a well-known sort of mantra—but also to which are added renew and recreate. "Renew" is about our human connection (affective/spiritual) to nature, and "re-create" is about the healing of nature and rebuilding of biodiversity.

The important message about the precarious state of wetlands preservation around the world and the Ramsar Convention on Wetlands of International Importance are emphasized at several locations throughout the site. Instruction is provided on the various threats to wetlands including drainage, dredging, damming, landfilling, pollution, and over exploitation. Ways in which personal action can help save wetlands are presented: save water, reduce and recycle, create a wildlife pond, remember that the inside drain connects to the outside world, etc.

The London Wetland Centre is interested in the power of people's unconscious and conscious connections with nature, and believes that in a way this represents some form of cultural heritage, the oldest and most profound that actually exists. In this respect, one of the most compelling missions of the Wetland Centre is to reconnect people with the past in remembrance that London, just like Paris, Boston, New York, Rome and many other major metropolises around the world, was at one time a "wetland city" (the Houses of Parliament further down the Thames, for example, were built right atop a large, then unoccupied, wetland). One way to accomplish this is to make people come to realize that nature is not somewhere else removed from the city; in other words, "nature" and "culture" are not by any means mutually exclusive concepts. The goal is to give people the key to the door of understanding that in this world of short attention spans and inundation by constant stimulation, that there is another way to look at the world around them. "Natural London" exists not only in terms of wildlife but in terms of landscape as well. With this purpose in mind, the grazing marshes at the London Wetland Centre call to mind an earlier agrarian peri-London at the same time as being situated within a late twentieth century context of of the

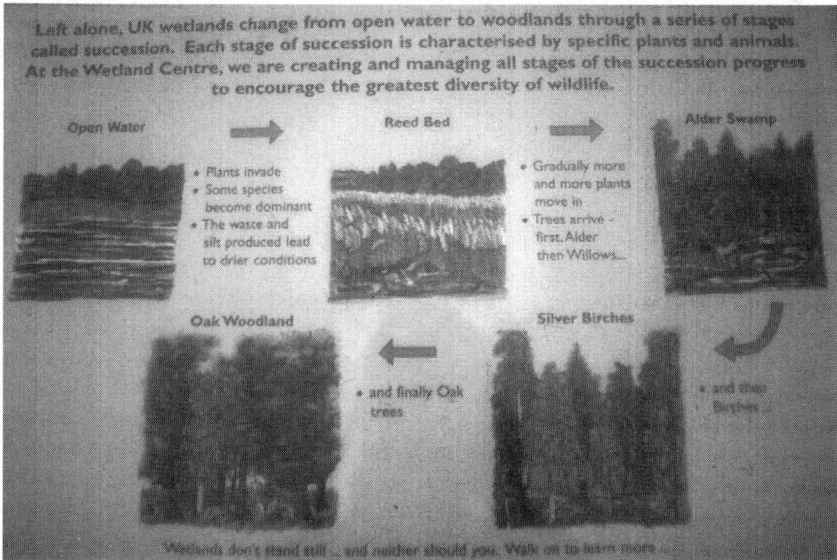

Figure 7.54 Example of educational signs explaining ecological principles.

whole as a sort of palimpsest of layering. Such an objective is very much within the spirit of Peter Ackroyd's assertion in his biography of London that London is a place where the past and the present combine all the time in an "echoic resonance."

In the Wetland Living exhibit, part of the Waterlife portion of the site, traditional artifacts, products and crafts inform about the long relationship

Figure 7.54 (continued)

Figure 7.55 Thatched roofed demonstration home in the Waterlife exhibit.

of human habitation and sustenance associated with wetlands. Objects here include fishing accessories, medicinal plants, food plants, building materials, transport punts, and duck decoys, all included within a thatched home (Figure 7.55). Other exhibits in Waterlife include the Pond Zone building loosely based on a Neolithic roundhouse and filled with images of a giant heron head, etc. to try to scale people down in size so the experience becomes a sort of magical place designed to compliment the real nature outside. The intent is to introduce the hidden drama of the garden or village pond designed in the style of an African game park with, for example, the giant underwater "lions" – voracious dragonfly nymphs, etc. The Duck Tales exhibit is a farmyard, complete with a Cotswold dry stone wall and friendly geese, that enables children to experience close encounters with one of the world's earliest domesticated animals. The WWT also commissioned leading gardeners to design a set of sustainable gardens such as the rotting log garden for insects and hibernating amphibians and mammals, a wacky water recycling garden that showcases drought-resistant plants and underground storage, and a land art garden to attract butterflies and other wildlife (Figure 7.56). In addition, in order not to intimidate normal gardeners with such avant-garde designer artistry, a much more humble area suggests what can be accomplished on small scales in a simple yet sustainable home garden. The messages imparted here are how to make a compost pile from household waste,

Figure 7.56 Sustainable gardens in the Waterlife exhibit.

Figure 7.56 (continued)

build mulch from chipped bark and grass cuttings, build a wildlife "hotel" from fallen leaves, branches and logs, and other simple energy efficient activities.

Walking through the Reserve portion of the site, visitors come across interpretive signs designed to resemble a set of "field notes" mislaid by some absent-minded naturalist. Topics include jottings and doodles on breeding behavior or the feeding habits of different birds. More in-depth interpretation is offered on touchscreens scattered throughout the site that inform about the birds' natural habitats, threats to these environments, and research that the WWT is doing relating to that particular species or habitat (Figure 7.57).

The London Wetland Centre also supports environmental education through providing a base for graduate student research. To date, theses arising from this university research have focused on such topics as wetland biogeochemistry, algal and macrophyte nutrient competition, the influence of avian herbivory, zooplankton dynamics, and the establishment of grazing marshes. Other research conducted by professors from universities throughout the UK has investigated fish dynamics, phytoplankton ecology, mammal abundance patterns, and released water vole population dynamics. Several regional school groups have developed ecology programs based at the London Wetland Centre. The Centre has developed liaison partnerships with various environmental organizations such as the Royal Botanic Gardens at Kew, the Royal Society for the Protection

Figure 7.57 Information touchscreens located out in the Reserve Habitats area.

of Birds, the BBC Natural History Unit, the British Trust for Ornithology, British Waterways, English Nature, the Natural History Museum, the Thames Estuary Partnership, the Federal Environment Agency, and of course Thames Water. In addition, monitoring and conservation action planning relationships have been established with a suite of local London biodiversity groups having particular interests in bats, wetland birds, butterflies and moths, dragon- and damselflies, and plants.

At the Scott Visitor Centre, the big observatory is designed to give people the "wow factor" right at start. Its two-story airport terminal window is deliberately designed to resemble an arrivals and departures lounge in what is supposed to be London's first "airport for birds." (Figures 7.58, 7.45). Screens depict different bird migration routes with touchscreens indicating the seasonal times of bird arrivals, departures and customs just like those for humans at nearby Heathrow (Figure 7.59). A mix of low- and high-tech interpretation material fills the Discovery Centre in an imaginative display of environmental education (Figure 7.60). Visitors move through a UK peat bog with giant sundew plants towering over them, before entering a mangrove swamp filled with giant models of mudskippers and fiddler crabs. Then visitors pass underwater through a Bahamian coral reef and into an Amazonian flooded forest before emerging into the "wetlands games arcade" filled with biodiversity-themed interactive computers automatically linked to other wetland conservation sites around the world. There is nowhere to escape the education

Figure 7.58 Early schematic drawing of the observation window at the Visitor Centre.

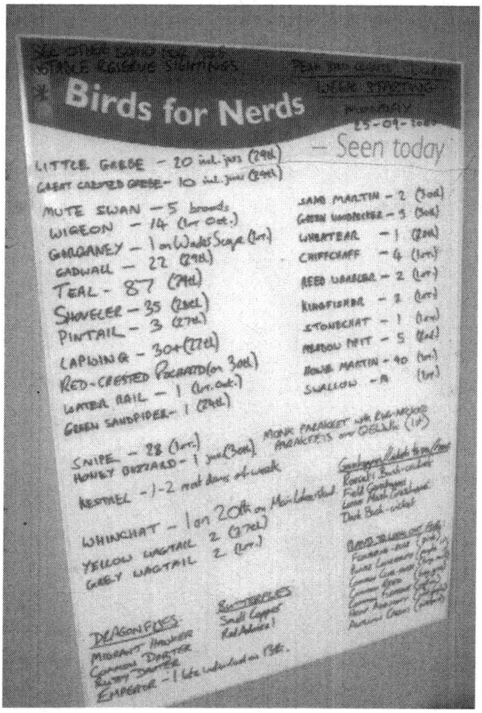

Figure 7.59 Bird departures and arrivals and other information boards at the "airport for birds."

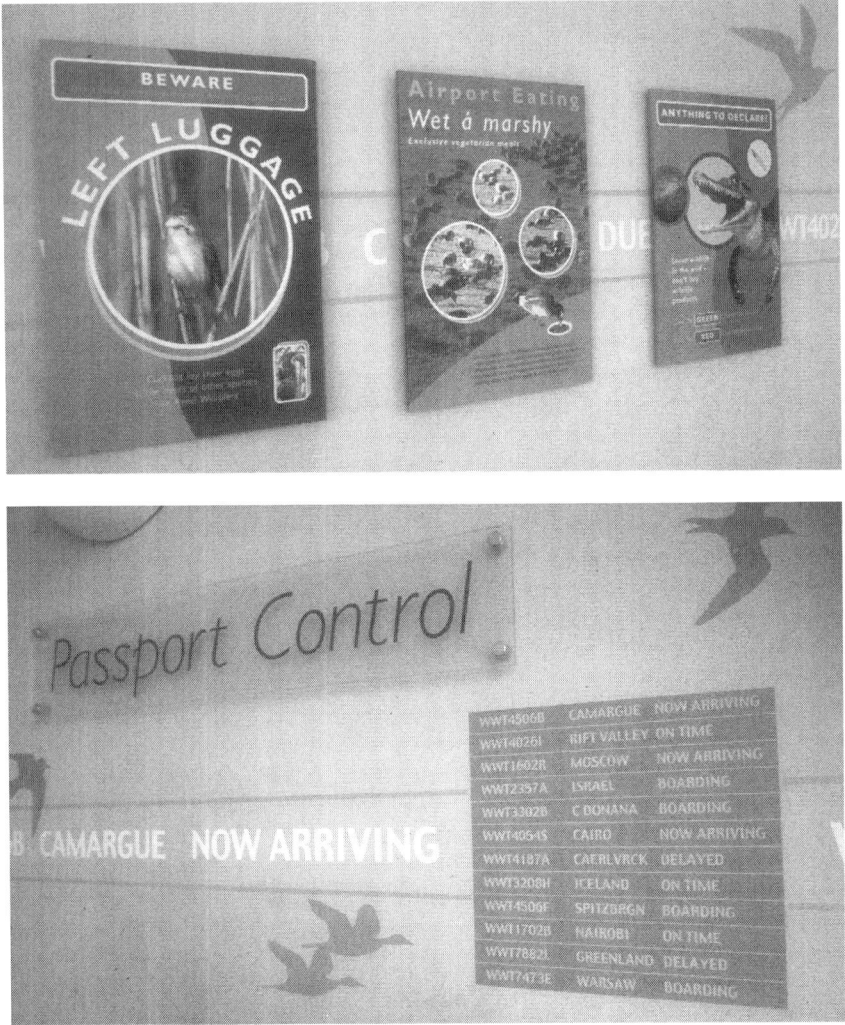

Figure 7.59 (continued)

message; even the toilet doors have messages instructing about how humans alter the water cycle. It is also of interest to note that because the place is managed as a visitor attraction, the site operations manager purposely is not a biologist but rather is a former manager of Madame Tussauds, a company that operates many visitor attractions including the famous wax museum.

The Shop has books, artworks, crafts, toys, and clothes (Figure 7.61) following a wetlands theme and fair trade ethics. The Art Gallery hosts painting and sculpture exhibits that change every two months and has

Figure 7.59 (continued)

Artist's impression of the Discovery Centre.

Figure 7.60 Early schematic drawing of the Discovery Centre.

included a Peter Scott retrospective and abstract paintings by local artists of the Wetland Centre itself. The Water's Edge Room is extremely popular and is used for functions and meetings by NGOs, corporations and individuals. Up to 30 percent of the total income for the London Wetland Centre now comes from rental of these facilities and affiliated catering.

A very busy full events program exists for the site. Daily events include volunteer-led guided tours for the general public, warden-led outings for birders, and pond dipping for any interested individuals. Special events have included themed weekends (on bats, otters, World Wetland Day, alternative therapy, holiday birding competitions, etc.). There are always numerous talks taking place as well as games and activities for children and craft workshops (e.g., willow weaving and

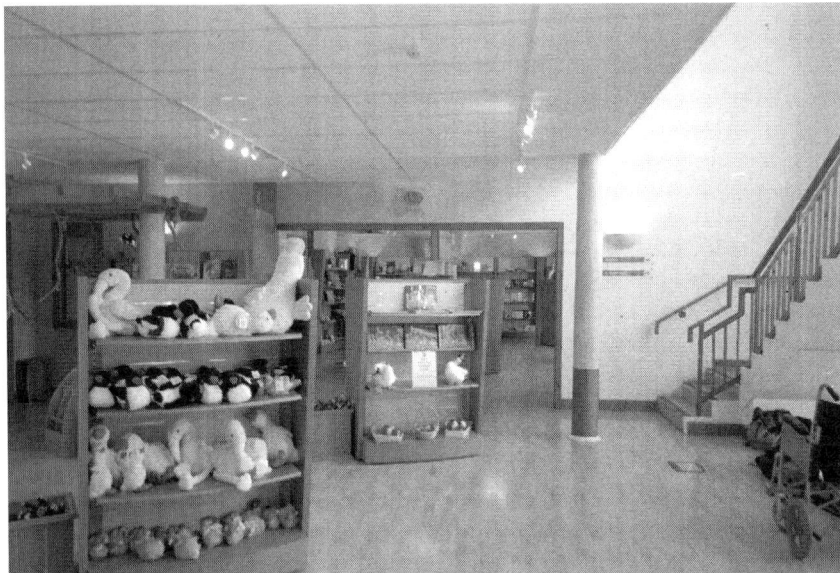

Figure 7.61 Gift shop and items available for purchase.

sculpting, painting and photography) for adults. Educational support materials (like teacher packets) are now delivered online.

During 2000, in the first year of operation, 150,000 people visited the Wetland Centre. Although early marketing research had suggested that as many as 300,000 would come, attendance is still on a growth curve and is currently (2011) around 200,000. Also, the London Wetland Centre has played a major role in recruiting new members to the national WWT. Since the Centre opened, membership has risen from 70,000 to over 100,000, of which 20,000 signed up as a result of visiting the London Wetland Centre.

Many people are making repeat visits, using the site not only as a wildlife viewing area, but also as a place to relax, as a play school, and as a location for lunch; in other words, simply as an urban park for recreation. Surveys indicate that visitor expectations are exceeded and that the interpretation duties have proven very successful despite the interesting observation that the audience is already quite educated when they arrive.

In terms of the local community, there are 140 volunteers who help not only with the periodic planting, but also with guided tours as well as with marketing and funding solicitation. In fact, there is actually a waiting list of potential volunteer wannabes. Also, by being situated where it is, the London Wetland Centre benefits from the attention and even help of many famous or well-placed neighbors.

Finally, if a major role in environmental education is to be a leader by example, an inspiring mentor, and a trainer for others to make the message afar, WWT's London Wetland Centre must receive glowing praise: "A magnetic attraction with an impressive underlying vision," stated the British Environment Minister; "An astonishing and miraculous opportunity to bring wildlife and wild birds into the centre of our great city, that must be the envy of Europe," said Sir David Attenborough; and "Its productive use of derelict land, enhancing biodiversity and promoting sustainable transport policies...is a superb example of collaboration between business and conservation," touted the leader of the Business and Environment Organization. One visitor, a landscape architecture student who spent six days at the site "watching, talking, recording...and getting lost in the labyrinth of wetlands," likened the experience to visiting a large Japanese garden, the act of staring out of a wildlife hide into the Reserve (Figure 7.62) being reminiscent of gazing out of a Japanese temple or moon-gate into a garden. Here the outside and inside worlds of nature and culture become seamlessly blended, boundaries cast aside, the visitor coming to occupy and submit him/herself to the "lure of the in-between." It comes as no surprise, then, that the site is now attracting an increasing amount of international interest, visitors and praise, such that the development arm of WWT is working on creating analogous centers in Hong Kong, Ghana, Korea and elsewhere, as well as increasing its general global outreach programs. And in 2001, the Wetland Centre won the British

Figure 7.62 The view from observation hides and blinds to the wildlife wetlands at the London Wetland Centre.

Airways Tourism for Tomorrow Award, the Palme d'Or as it were of the sustainable ecotourism world, usually reserved for some naturally preserved park located in a romantic and exotic tropical corner of the world.

Future plans and summary

As Britain's urban centers continue to develop and sprawl into the countryside people are expressing a growing desire to restore their links to wildlife. Urban dwellers crave quiet wild spaces where they can find solace and escape from the hurried din of city life. They need "urban wilds," pockets of countryside existing within the city where wildlife can live in safety. Instilling such a feeling has been a major task of the London Wetland Centre. Further, the Centre has proudly carried forth the WWT's major objectives of conservation, research, education, and recreation. Given the success of Phase One, plans exist for developing levels Two to Four. One of the plans calls for creation of a Tropical Wetlands exhibit (Figure 7.63) patterned on the successful Eden Project of greenhouses and botanic gardens located in Cornwall. Another future project in the planning stage is for a Thames River Life exhibit focusing on river life (Figure 7.64). An architect known for his organic-inspired designs has been commissioned to design the building as a sort of aquarium turned inside out. Plans also exist for constructing a children's play area to encourage learning through play.

In the end, has it all been worth it? Has the London Wetland Centre been effective? If one measure of judging this is its role in environmental improvement, the site itself can most definitely be regarded as one of the world's highest-visibility success stories of urban wild renewal. Equally, if possibly even more significant, is the fact that influence of this premier environmental education center transcends the physical bounds of its site through global outreach. For example, the WWT brought the mayor of the province next to Seoul to see the London Wetland Centre. This politician's home is located beside the most important mudflats in the entire Asian flyway, which unfortunately were being destroyed at an alarming rate. After only five minutes beside the Wader Scrape at the London Wetland Centre, and receiving an on-the-spot lecture about the importance of mud, the mayor returned to Korea and immediately dedicated 80 square kilometers of mudflat wetlands as a protected area. So in this respect, if the WWT has done nothing else with the £16 million that it took to build the Centre than the education of this single influential politician, it has been worth it in terms of global biodiversity protection. The London Wetland Centre, therefore, is already functioning as a fitting testimonial to the inspired legacy of Sir Peter Scott, one of the world leaders in avian conservation biology (Figure 7.65). Although Scott was not to live to see the opening of the Wetland Centre, his last painting of birds in flight over this section of the Thames was his own artistic vision for what the site could, and has, become.

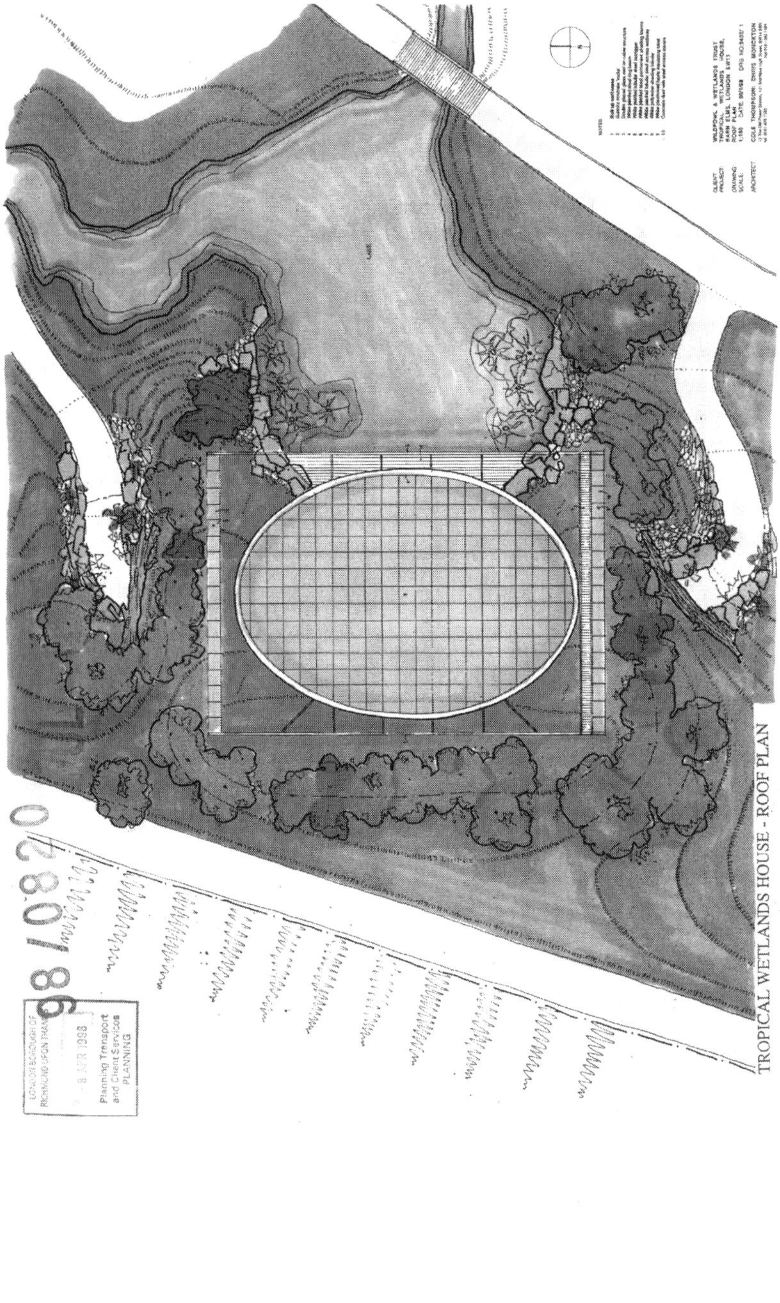

Figure 7.63 Architectural and landscape plans for the future Tropical Wetlands Exhibit.

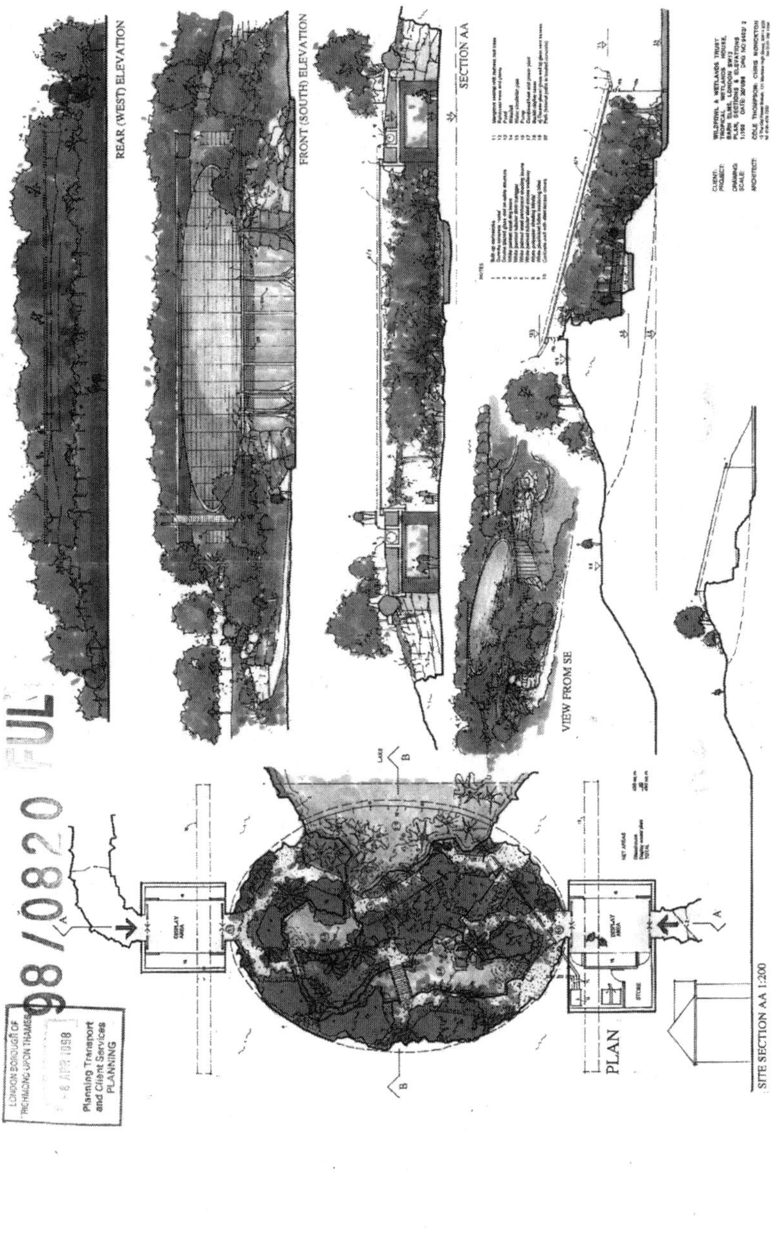

Figure 7.63 (continued)

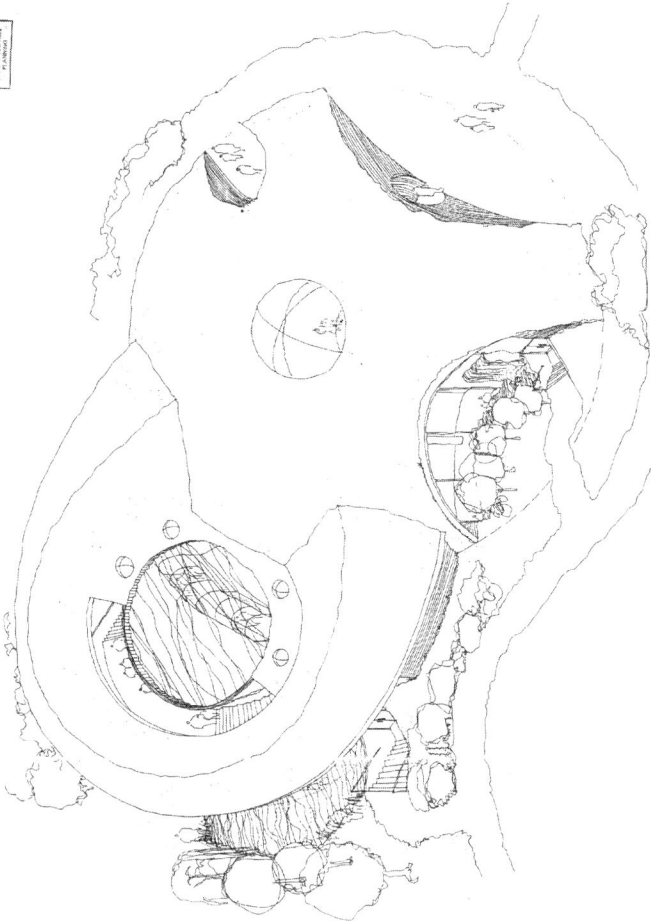

Figure 7.64 Architectural drawings for the future Thames River Story exhibit.

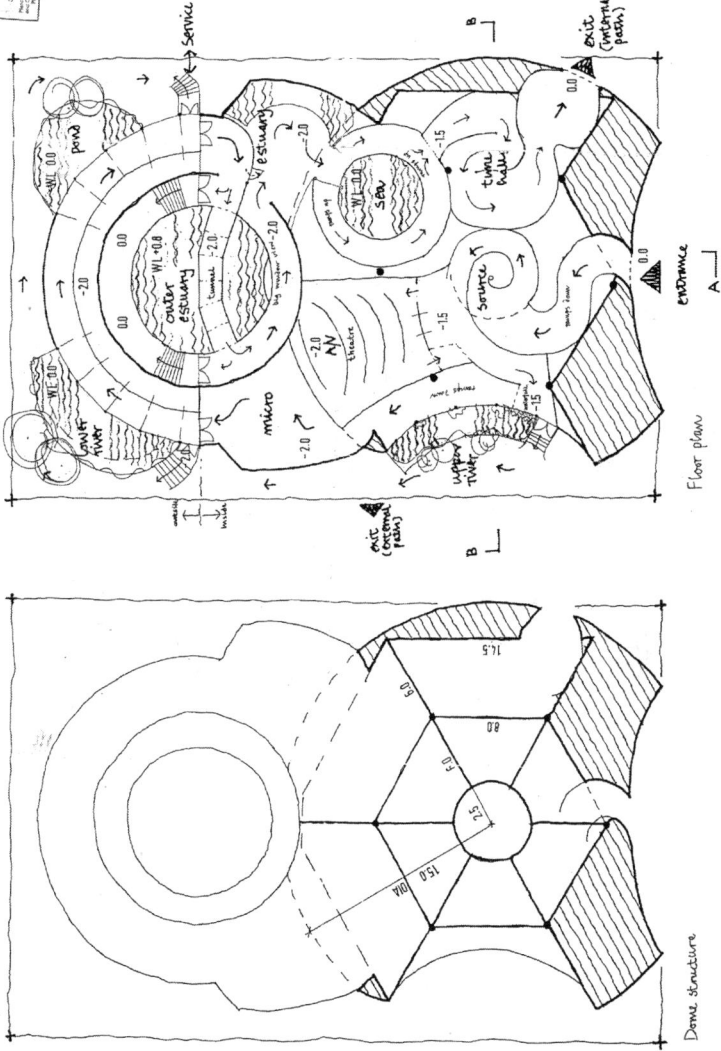

Figure 7.64 (continued)

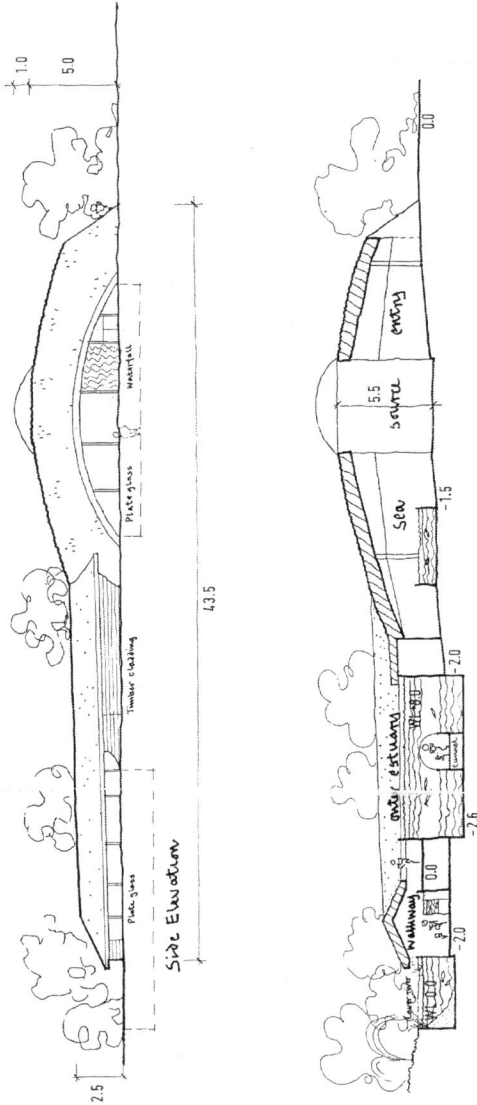

Figure 7.64 (continued)

Figure 7.65 Statue in front of the London Wetland Centre of Sir Peter Scott, founder of the Wildfowl & Wetlands Trust and the World Wildlife Fund.

Sources and references

Ackroyd, P. 2000. *London, the biography*. London: Chatto and Windus.

Anon. 1987. *Barn Elms. A vision for the future*. Thames Water and the Wildfowl & Wetlands Trust.

Anon. 1987. *Barn Elms Reservoirs. Technical appendix 4: Landscape and visual Appraisal*. RPS Environmental Planning Consultants Ltd.

Anon. 1989. *Barn Elms Wildfowl & Wetlands Trust Centre*. Thames Water and Wildfowl & Wetlands Trust (WWT).

Anon. 1989. *A Wildfowl and Wetlands Trust Centre at Barn Elms. A Proposal for Thames River Authority*.

Anon. 1992. *Barn Elms*. News update flyer, WWT.

Anon. 1997. *Concept to creation: Business enables wildlife to flourish*. Wetland Center Business, WWT.

Anon. 1999. *Wet, wild and wacky: Interpretation at the Wetland Centre*. The Wetland Centre Newsletter, WWT.

Anon. 1999. *Environmental Review*. Thames Water.

Anon. 2000. *Wetland Centre monitoring report, 20th progress report*. July–September, WWT.

Anon. 2000. *Less than 100 days—the countdown to opening begins!* The Wetland Centre Newsletter.

Anon. 2000. *Let the show begin!* The Wetland Centre Newsletter.

Anon. 2000. *The Wetland Centre information brochure*. WWT.

Anon. 2001. *The Wetland Centre guide*. WWT.

Anon. 2001. Wetland winner. *The Guardian*. 6 October.

Anon. 2008-10. The Wetland Centre. www.wetlandcentre.org.uk

France, R. 2003. *Wetland design: Principles and practices for landscape architects and land-use planners*. New York: W.W. Norton.

Peberdy, K. 1997. The making of a wetland. *Wildfowl & Wetlands*, 124:16–19.

Peberdy, K. 1999. Planting a wetland wonder. *Wildfowl & Wetlands*, 127:16–19.

Peberdy, K. 2000. London life just got wilder! *Wildfowl & Wetlands*, 132:17–21.

Wetland Link International. 2006. *Developing a wetland centre: An introductory manual*. WLI.

Whitehead, M., and D. Hulyer. 2000. London's new wetland centre—where Conway's bullfrog meets zoo 200. *Internat. Zoo News* 47/7:419–25.

Winter, S. 2001. Wetland creation along the Thames: Boundaries and the lure in between. Penny White Award Project. Harvard Graduate School of Design.

chapter 8

London case study questions

Answers by Malcolm Whitehead, Kevin Perbedy, and Doug Hulyer

How does the balance between the natural wetland areas, important presumably for sustaining native biodiversity, and the obviously artificial, zoo-like wetland areas, important presumably for sustaining revenue, work? Are most of the visitors to the Centre going there for the zoo or for the more natural areas? And what sort of responses have you received from naturalists?

World Wetlands—the captive area with naturalistic, fabricated tableaux—is physically separated from the wild, created habitats. Birders and general naturalists prefer the "wild" areas because they are so biodiverse. Children like the proximity of the "zoo." Most people like the blend of experiences...but they keep coming back to the wild areas.

What about the increase of birds at the Wetland Centre and proximity to the airport?

There were no deleterious effects from the airport and we were actually more worried about plans for an adjacent sports field and the ensuing light pollution, but increases in waterfowl abundance still occur.

What about the reciprocal in terms of how this increase in birds might pose a threat to airplane safety at Heathrow?

Birds are pretty low on the list of perceived threats to Heathrow. There were no objections from the Civil Aviation Authority, and we don't consider it a problem.

What sort of ornithological research is being conducted in the natural areas? Have there been any interesting interactions between the native birds and the exotics?

The research is mainly involved in monitoring the increases in species number and overall population trends. After the first 5 years of collected data there was enough evidence to have the site recommended as a UK Site of Special Interest (SSSI). Research studies have examined the interactions of macrophytes and birds as well as the establishment of the grazing marsh. Water quality is actively monitored and the population dynamics of introduced, captive-reared water voles are closely studied. Other work studies the competition between fish and birds.

In terms of interactions between native birds and exotics (the latter are physically separate in the World Wetlands area—although wild birds can fly in), it is mainly predation of native birds upon the exotics. There was one heron in particular who apparently specialized in eating chicks of one of the world's rarest waterfowl species. Gulls and crows have to be watched and rats have discovered the site. Pigeons eat captive bird food and are potential disease carriers for other birds, so we have to be vigilant.

How many clipped-winged exotic bird species are there in the World Wetlands exhibit? How were they collected? And have there been any problems with feral cats, etc.?

There are about forty-five species and they are kept at low stocking levels to preserve water quality and vegetation. It should be noted that the WWT is probably the world's foremost exponent of captive waterfowl conservation with over a half century of expertise in breeding ducks, geese, swans, and flamingos. Several species, such as the Hawaiian goose for example, owe their continued existence to WWT captive breeding programs. We have a dedicated captive breeding (ex-situ conservation) officer at our headquarters in Slimbridge and actively participate in many regional and global captive conservation programs. WWT maintains close links with some of the world's best zoos and related institutions, as well as many government departments of conservation.

The captive World Wetlands area is surrounded by a fence to keep out predators such as foxes and cats which do, however, take small numbers of birds and water voles in the other wild areas of the Centre. Of greater danger are occasional flocks of Canada geese which can trample and consume a wetland in a short time.

So much of the focus appears to have been directed toward birds in the original spirit perhaps of the WWT, but some mention was made about smaller habitats constructed for amphibians and for aquatic insects. Could these be described in a little more detail?

Protective hibernacula have been created for amphibians. These are small (0.5 m tall by 2 m long) earth mounds under rocks where the amphibians can overwinter. Ponds have been created especially for amphibians and dragonflies that include diverse plant communities and sheltered areas with haul-out basking stones. Dragonflies need emergent plants to allow for their own emergence into adults. Some ponds are regularly drained to remove fish predators.

What is the vegetation maintenance plan? And who covers the costs?

We did build in physical features into the site design to stop emergent plants from spreading and choking the central Main Lake. These were a deepwater 3-m perimeter channel that operates like a fence and for each reed bed subbasin, a freeboard of about 1 meter for water regulation. This latter took the form of raising the water level as reed detritus accumulates until the point is reached where you can't stay ahead of the process at which time rotational cutting is required.

Management is funded from gate revenues. It is important to realize that the London Wetland Centre is a paid attraction that is run as a business integrated into the overall budget for all nine WWT sites around the country.

What then is the number of staff at the London Wetland Centre itself?

About forty full-time staff work at the Centre with additional seasonal hirings for grounds and restaurant work.

Are animals used in the grazing meadows?

Initial plans in this regard were temporarily halted because of the nationwide outbreak of hoof and mouth disease. In the long-term, sheep are being used on the dry grasslands and a rare breed of cattle (such as Dexters) on the grazing marsh.

Why were you so constrained by importing and exporting materials? And would the final design have been different if this had not been the case?

It was a planning precondition based on wanting to restrict the movement of vehicles in the neighborhood. About 500,000 cubic meters of spoil were shuffled about from place to place, so given the number of lorry movements that would have been required, it would have never been really feasible to remove all the material. At one stage, we had planned to use barges along the river, but the idea was shelved by an archeologist who found remains of an ancient oak woodland along the entire shore which had not been known previously. As a result, we weren't allowed to build a pontoon and so lost the chance of taking any materials off site.

Since this limitation was determined very early in the process it became the operating condition so it is difficult to say how the resulting design might have changed had this constraint not existed. On the other hand, it was always our desire to recycle all the materials, especially the concrete, and so we are glad this was accomplished because it became one of the most positive parts of the project. The overflow carpark, for example, with the recycled crushed concrete turns out to perform better than new commercial "grasscrete." In short, the pressing desire for sustainable design in terms of material recycling probably meant that the final design would have been little changed with the removal of this "constraint," despite the fact that it would have been much cheaper to simply throw it away.

With so much attention to constructing soils and establishing plant communities, was use made of professional soil scientists and botanists?

Detailed analysis of the available soils as a planting medium (in addition to construction properties) was a major part of the initial site surveys. WWT's own botanists and ecologists worked with soil scientists to establish the ideal soil combinations for different plant communities under various hydrological conditions.

What was the overall budget?

It is really difficult to say because a large proportion of the work was involved with taking the Barn Elms Reservoirs out of service and this was paid by Thames Water. The main earth-moving contract that included building the subbase and footpath and foundations of the hides and other basic land-forming came to about £5 million. In the end, it cost £16 million in total, going from nothing to a fully-opened attraction, with much of the £5 million that was raised later being directed to outfitting the buildings with the exhibits and those in the grounds.

What is the general quality of the Thames River water supply? And were/are there any forested communities present on the site?

We do have to rely on raw Thames water, which has been improving over the years but which still contains elevated nutrient levels—phosphorus in particular—that are still a problem and were always recognized to be such. We have built in buffer systems such as the reed beds which are particularly structured to filter and treat the water before it goes into the most sensitive ponds, which contain the most dragonflies and amphibians. But given that most wetlands are not particularly effective at phosphorus removal, we are building treatment filters on all of the water-control structures by using calcium-rich substrates to help strip the nutrient from the water. Also, we plan to try other techniques, such as using oysters or mussels in cages to try to remove even more phosphorus from the water. The large mid-summer algal bloom problems have declined over time. We might also experiment with developing new techniques of using barley straw filters that actually release an algicide. All of these approaches do not get to the root of the phosphorus problem but do alleviate the consequent algal problem. Monitoring studies have found that the reed bed system does successfully take out most other contaminants (up to 95 percent), though after large storms we do sometimes get an inflowing pulse of heavy metals because of street runoff entering the Thames.

There are wet ecotonal areas with most of the banks planted with a mixed woodland progressing from a wetland woodland dominated by alder, willows, and downy birch, to a dry woodland dominated by ash, field maple, birch, and ultimately oak on some of the banks. About twenty-five thousand trees have been planted over the entire site.

How much water is pumped from the Thames through the system? And where precisely does it originate; in other words, isn't the river still tidal in Hammersmith? In short, how does the hydrology operate?

Water comes from the nontidal part of the river at Hampton, about 15 km away. It is carried along a Victorian era pipeline with a small amount being pumped off at the Wetland Centre while the majority of the flow is transported to northeast London. Water flows from the northwest to the southeast across the waterbodies of the Centre before ultimately being returned to the Thames. When the pumps in northeast London are switched on, a fall in the water head at Barnes means that pumping is needed into the London Wetland Centre. At other times, as the water level in the Centre is lower than at the intake at Hampton, flow is by gravity. A large recirculation pump also exists at the Centre to keep water moving from cell to cell.

How has Thames Water reacted to the project? And what, if any, is their ongoing relationship with the Wetland Centre?

Thames Water is very supportive of WWT's efforts. It is their land after all. We maintain a loose, informal contact. Some of their staff are seen at various biodiversity and education meetings.

What was the rationale for the housing development, and were any ecological designs used?

The initial rationale was purely economic, so we looked at the development through real estate eyes as a vacant plot. But as an organization dedicated to sustainable development, we searched for a developer who would take on all those sustainablity issues—water runoff, sustainable use of building materials, etc. To be honest, though, the developer was not really there at the start, but since that time they have learned enormously about these issues and are now sponsoring environmentally friendly housing elsewhere in the country because they can see the arising commercial benefits. There are three wildlife corridors that extend back through the development and form some of the backyards, and the pond which we think was an old WWII bomb crater is now linked to the Wetland Centre such that we definitely have evidence of bats feeding and moving between the two locations, as well as migrating amphibians. The Centre was planned and built at a time when the WWT was just starting to look upon wetlands as utilities for sewage treatment. So for the more recent center built out in the country at Slimbridge all the waste from the visitor center there will go through a treatment wetland. However, we couldn't imagine trying to do this in London with such untested technology at that time. Plus the physical space was simply not available.

There seems to have been a missed opportunity in green building design by not having wetlands treat the toilet and graywater from the current facilities' building site.

That's where the new Thames River Life aquarium will be going in, and because any aquarium is a high-energy, high-water-use facility, and because we always believe in balancing the educational learning element with the sustainability element (i.e., practicing what we preach), we are seriously thinking about such things. It is important to realize that the overall site design isn't finished yet. We first needed to open, get people through the door, and test out ecology and education, as well as economics. In a sense, we look upon the entire site as one big experiment in sustainable development, and we are by no means finished yet.

Still, it must be admitted that despite the words about the five secret functions of wetlands, it could be argued that a disproportionate emphasis is given to wetlands as glamorous habitats for birds and little else.

We don't glamorize wetlands as only oases for birds. We also address other wildlife issues as well as promoting wetlands as cultural resources and providing a theater for connection to nature.

How has the existing community of abutters reacted to the Centre? Have local businesses prospered? Do pints cost more in the local pub? Have housing prices and/or property taxes risen? Are the neighbors peeved at the increase in car traffic from visitors, etc.?

The *Sunday Times* Property Supplement estimated that proximity to the London Wetland Centre had raised house prices by 3 percent. Neighborhood pints do cost more but that is because of the level of affluence in the area. Most neighbors love the Centre; after all, they could have ended up with a shopping mall next door!

It seems the project is largely focused on concepts of restoring London's natural heritage almost at the expense of its cultural heritage. What discussions took place, if any, about the idea of preserving some of the site's industrial heritage as some sort of park along the lines of Gasworks Park—which has come to define public space in Seattle—rather than merely wiping the slate clean? And as perhaps a poor compensation, are there any plans to host displays about the industrial history of the site in relation to greater London's network of waterworks, etc.?

Onsite touchscreens do allude to the site's history and temporary exhibits and talks are planned about this issue. But the overriding objectives were and are unapologetically about rebuilding biodiversity and connection to nature.

This is a very impressive accomplishment, so much so that it is worth rearranging your flight through Heathrow and go for a 4- to 5-hour visit. What is most important is that it is a site that is so easy for anyone to get to with there being no exclusion based on car ownership. And though there is quite often a distance between quality design and ecological concerns, with the exception of the Visitor Centre which some might find to be slightly disappointing, the other buildings that do the interpretive job are sensitively done, well-sited, and the place really works. There is an awful lot of interpretation, as for example the computers that talk to you in the middle of the wetland, which some may find a bit difficult to take. So perhaps there is a need for some more visitor-use surveys to find out how people like being talked to in the middle of a marsh by a machine.

As part of planning, some of the money (half a million pounds) from sales of the houses went toward support for sustainable transport. The London Transport Authority started a "duck bus" for transporting passengers from the Hammersmith Tube Station and directed lots of money at marketing on the Tube. Unfortunately, within 2 weeks of opening, the IRA planted a bomb which exploded on Hammersmith Bridge, which is the main way of getting from the London Tube across the river to the London Wetland Centre in Barnes.

We have sixteen touchscreen computers onsite. Those situated inside the Visitor Centre function as an interactive interpretive medium; for example, "what's that bird?" comes up and you touch a silhouette that then asks about two flight patterns, etc. So in a way it functions like an electronic identification key. But we try to zone our interpretation facilities on a gradation in the field such that they are not present in more natural areas but restricted to the core areas. And yes, some people have complained that they are too loud, so we have dropped the volume. A survey of the interpretation services conducted by the University of West England showed that the user satisfaction rate was very high with visits recorded of over 2 minutes at the touchscreens. And although we cannot separate out the individual bits, what visitors tell us they get out of the whole visit is that they are picking up the big message about wetlands; in other words, that such landscapes provide benefits to both birds and humans, that they are disappearing at an alarming rate, and that we have to do something to protect them, etc. What was not expected, however, was that the visitors seem to have a very high preknowledge before they come, so perhaps we have to do a better job of outreach rather than preaching to the converted.

How does the "fun" element of education come in?

Our signage has an element of whimsy and irreverence about it. In one case, for example, for the signboard where birders jot down the names of their latest spotted finds at the main observation "airport" window in the Visitor Centre; we may have gone a bit to far when we identified it as "Birds for Nerds." The largely white male, sometimes socially inept, birding zealots, failing to appreciate the humor, became upset so the sign moniker had to go.

Are any approaches being investigated to increase the diversity of visitors? How many of these are foreign tourists? And how have the London tourist authorities been marketing the Wetland Centre, if at all?

WWT actively fundraises to implement a social inclusion program to encourage visits by ethnic minorities, and people from deprived areas and socioeconomic groups. The project employs a Social Inclusion/Community Liaison Officer. We are also developing a London Community Wetlands Project to reach out across all of the capital city. Most of the visitors are local Londoners since the Centre is located slightly off the main overseas tourist hotspots circuit.

Are there any future plans to increase the visibility of the site in relation to the Thames River, which is currently ignored as an entrance point? In this light, are there plans to coordinate with the various Thames tour boats that work their way upstream to Hampton Court to arrange a docking drop-off point as is currently done for Kew Gardens?

Not yet, but we're very receptive to the idea notwithstanding some concerns about access to the archeological area mentioned above.

What about the referred-to "problems" associated with the Visitor Centre?

Most of the Visitor Centre is very successful, especially the "airport" observatory, shop, restaurant, gallery, and theater. The Discovery Centre is popular, but we have some misgivings about the clarity of the delivery of messages and the quality of some interactive elements. Redesigns exist on paper which will be implemented when funding is obtained.

What are your future plans for managing the vegetation, as for example, problems with colonizing exotics?

Rotational cutting of emergents will occur, but there also will be a need to physically go out and pull out seedlings at a nonsensitive time when there are not a lot of birds present to disturb. We did recognize at the start that it would be quite a difficult management task to sustain our planting diversity, so we use about thirty volunteers a week to do groundwork in support of the four full-time professional groundskeepers. We have been invaded already by three nasty alien plants, one of which—floating pennywort which grows over open water and if you break it up the particles just float downstream and reroot— required use of herbicide, which was something we had said we would never do. For this particular plant, the real way to get at the problem is to control it at the source, and we now have been successful in restricting it to a single pond.

Is there education about the natural process of succession, and how does this fit into the issue of management?

One way to set the successional clock is to manipulate the water levels. A recirculation system exists because there is a risk that the piped water supply might not always be there. So we periodically drain the water bodies down. There is also a problem with fish populations which is common in closed lakes where they can begin to compete with birds. As a result, we have to make value judgments if we plan to de-water an entire lake to remove the fish. In terms of explaining what succession is, yes we do have a Succession Trail with a group of ponds set at different temporal stages with associated interpretation which we expect will have to be altered as the systems mature. We are now at the stage where we do need a regular but flexible management plan following the first 3 years of largely trial and error. As such, we may need to decrease proportional use of volunteers in favor of professionals guided by more rigorous quality standards. Also, a good deal of our overall education message is telling people that nature is not static and dealing with questions of manipulation, when to play God, and the like. In a sense, we plan to use these management "problems" we are wrestling with as an opportunity to bring these issues into a wider public debate with informed laypersons.

How does the London Wetland Centre fit into the larger landscape in terms for, example, the Thames River Walk?

For birds, it is an increasingly known landmark—"those nice people at WWT opened a five-star hotel, resort, and restaurant complex just for us"—and it's easy to find, being right along the Thames flypath. The Centre has also penetrated Londoners' mental map of new landscapes and places to go for a visit.

What, finally, are the most important lessons that are being adopted to the WWT's work in creating similar centers elsewhere around the world? And given the somewhat eccentric trait of the English for bird-, train-, and plane-spotting, how successful are such projects expected to become in countries where this tradition is neither as strong nor perhaps might even exist?

There are thousands of technical lessons relating to water quality, soil nutrient status, alien species management, environmental monitoring (essential for habitat management), good project design and management, and the overriding sheer expense of habitat creation on such large spatial scale. And of course we have learned from our mistakes about all the above and more; in other words, so as not to emulate the late English wit Peter Cook who when asked if he had learned from his own mistakes, joked "Yes, I'm sure I could repeat them exactly." Other lessons include:

- Each case is different. Wetland centers need to be customized; it isn't appropriate to transport the London model as a clone elsewhere, no matter what the wishes of its many fans might be.
- Listen.
- Hire the best.
- Plan to fail.
- Remember that the problem with communication is the illusion that you've achieved it.
- Be people-focused. Start where your audiences are. Not everyone is a biologist, landscape architect, or even shares the same beliefs, incredible as that may seem.
- Integrate natural/biological and cultural concepts and themes.
- Be interpretation-led, and don't skimp on any areas.
- Development people come from Mars; operations people come from Venus. Few combine both skillsets, so include both in your planning team.
- Build operational and maintenance costs into your development plans.

- Don't believe preopening forecasts from expensive marketers. A safe rule is to divide by three.
- In terms of "spotters," the ardent birders and such will always come regardless of the country. The English are not the only "mad" ones out there in this regard.
- Wetland centers are about much more than this, though. They are both biological and cultural phenomena. Wetlands supported and sometimes continue to support our great cities and civilizations. Connection to nature through wetlands as a therapeutic recharge will become increasingly important as more and more of us live in big cities. Wetland centers can be parks, meeting places, somewhere to fall in love, have a meal with friends and family, watch your children grow, etc. This is something that touches all global cultures and transcends simple bird watching tribal subcultures.

chapter 9

Clark County Wetlands Park and the London Wetland Centre integrative themes and lessons

Interdisciplinary work

Professional diversity

Successful regenerative landscape design for recreation and ecotourism is an extremely complex undertaking, transcending the knowledge base of any single discipline, and therefore necessitating the establishment of an interdisciplinary team from the start. At one time, the regeneration work at the London Wetland Centre involved over two dozen specialists onsite to implement the designs, and the incredibly comprehensive planning process for the Clark County Wetlands Park (almost unequalled anywhere in scope) involved numerous representatives from almost thirty different organizations/agencies/companies. In such situations, project leadership and management obviously becomes critical. This implementation strategy can be regarded as the "cathedral view" of project development, whereby each "stone mason" works away on a part toward a grand and unified whole. It is critically important, therefore, that those directing the overall work can deal with the idiosyncratic particulars of specific problem solving while always holding the broader vision firmly in place. Of those disciplines that integrate such holistic thinking with hands-on practical skills, landscape architects can often ably fill such leadership roles in addition to their oft-credited roles in establishing site identity, sense of place, and highlighting various features, etc.

Community networking

Regenerated sites need to be reclaimed in the consciousness of the rejuvenated neighborhoods just as much as they need to be physically remediated. As a result, it is essential that a strategy of networking be established early that brings on board local authorities involved with transportation utilities, recreation, etc., as well as neighborhood public interest groups.

Too often the problem with any development site, whether reclaimed brownfield or fresh greenfield, is that it is conceptually considered and then physically realized as being completely uncoupled from the surrounding landscape. People often regard the promise of regenearating postindustrial landscapes with suspicion, wondering why all those resources are to be spent on such an abused location that is unworthy of attention, much less care. Often then, one of the biggest challenges that planners and developers of regenerated sites have is to restore the promise of a recovered landscape in the mindset of the community that is used to looking upon the site with only derision. It becomes especially critical in such situations, as the London Wetland Centre found, to inform and engage the surrounding community from the very start as a strategy to bring them on board. The comprehensive outreach program employed during all stages of planning the Clark County Wetlands Park (including meetings, flyers, websites, etc.) is a model about how to empower the local community in decision making.

Project development

Precedent examination

It is human nature to try to give the impression that the wheel is being invented in each and every project being implemented. More rare is the situation in which project developers give credence to ideas that have been tried and proven elsewhere by others. Engaging in a background survey of how similar regeneration projects or other public parks have been planned, designed, implemented, and managed goes far toward establishing a credibility in the ensuing process of redeveloping and recovering the landscape under consideration. In this regard, by contexturally situating the Clark County Wetlands Park within a larger framework of parks in the American Southwest, the planning consortium was able to estimate development and maintenance costs as well as to circumvent many problems before they occurred.

Master planning

Regenerative landscape design for recreation and ecotourism is an incredibly complex and incompletely understood science or art endeavor. Given both the importance of the activity, as well as the frequent uncertainly of its execution, a comprehensive master plan should be regarded as de rigueur for site reclamation or regeneration. The planning process for transforming the degraded Las Vegas Wash into the Clark County Wetlands Park sets the standard for this procedure.

Site inventory

It is inconceivable than any viable process of ecological restoration or rehabilitation as part of a regenerative landscape design project can be initiated without a firm foundation of good scientific understanding of the existing ecological conditions. Although often shortchanged in terms of budget and time, it is imperative that a solid base be obtained upon which to affect landscape recovery. In this regard, the effort by the consortium involved with planning the Clark County Wetlands Park to obtain a survey of biogeochemical site conditions is to be strongly commended.

Land ownership and encroachments

Postindustrial brownfields and derelict graywaters are often surrounded by a complex mosaic of land ownership where individuals' interest in, and sometimes protection of, their property increases in parallel with the recovery of the neighboring abused or abandoned landscape. At the same time, many abutting property owners may either be knowingly or unknowingly encroaching upon what they had hitherto regarded as simply wasteland. The delicate but essential job of approaching abutters and of possibly acquiring desirable lots, such as undertaken in the Las Vegas Wash, for example, can be integral to sustaining the vision of the created park.

Real estate

Housing developments can be beneficial to implementing regenerative landscape design. Sometimes courage may be required to bring about resolution of such projects. The "integrated scheme" put forward by the Wildfowl & Wetlands Trust in terms of making the bold decision to relinquish about a quarter of their site for an "enabling development" of premier homes was the absolute key to the success of the London Wetland Centre, so much so that without the influx of the real estate revenue little would have been accomplished at all.

Opportunities

Imagination is required to turn what most perceive as problems (for example, brownfields) into opportunities for regenerative landscape design for recreation and ecotourism. Several, less imaginative groups turned down the site for the future London Wetland Centre simply because they could not see out of the box. The best regenerative landscape design projects are those in which the planners dream large dreams that can capitalize on the hidden potentials lying in wait in the postindustrial or postagricultural derelict landscapes.

Plants

All efforts should be made to use native local plants whenever possible to provide the best chances for regenerative success. All recovery landscapes go through a stage of "scruffy adolescence" during which plants gradually become established. The likelihood of "greening" and recovering damaged postindustrial or postagricultural derelict landscapes improves markedly with careful attention paid to the type and source of the native flora. In some situations foresight is needed, as the London Wetland Centre found, in terms of growing such vegetation well in advance of when it would be planted into the newly refurbished site. It may be important to plant mature native trees to kickstart the recovery process by getting the jump on the colonizing exotics that are often characteristic of damaged landscapes.

Temporal planning and management

The success of any regenerative landscape design project is a long-term investment that requires comprehensive maintenance and management, a plan of which should be set in place before the first bit of soil is turned over. Signature regeneration projects are by definition complex undertakings that require visionary management. As the London Wetland Centre demonstrated, this needs to be formally planned and budgeted right from the very start and never approached as an afterthought. Both the London Wetland Centre and the Clark County Wetlands Park established long-term goals that are significant in that they are truly decadal in their respective scopes. The consortium planning the future of the Clark County Wetlands Park used build-out models to forecast development changes as a means to ensure that their regeneration plans were not frozen in time. Another important lesson from the Clark County Wetlands Park case study is the demonstrated importance of tackling problems—such as ongoing erosion in their case—when they are first identified before those problems are allowed to worsen and therefore become much more expensive to fix at a later date.

Funding

Regenerative landscape design projects amount to little unless supported by adequate funding to implement the often lofty recreation and ecotourism goals. The number of "orphan" regenerative landscape design projects around the world is legion. All that wonderful design and attention will be for naught unless there is a funding base in place to sustain the vision through time. The London Wetland Centre approached this problem by actually hiring a professional fundraiser who, against the pundits' predictions, was able to bring in enough money to start the work and to keep it going in terms of maintenance into the future. Both the Master Plan and Adaptive Management Plan for the Clark County Wetlands Park and Las Vegas Wash spend a considerable amount of space on presenting clearly enunciated budgets for a spectrum of development activities ranging from operating the visitor center to construction of each trailhead entrance. The process is transparent and the public can monitor financial accountability at all stages.

Site construction and management

Demonstration projects

For large sites or large problems, as for example Clark County Wetlands Park with respect to removal of invasive plant species, there is a need to move regeneration forward through a step-wise, iterative strategy of concentrating on localized demonstration projects rather than attempting to undertake a broad-brush, and therefore superficial, repair of the entire landscape.

Phasing

A detailed phasing and staging plan is essential for managing and successfully completing complicated regenerative landscape design projects. The London Wetland Centre's "constraints" in having to keep all materials on-site necessitated an enormous orchestration of moving soil from place to place while the site was being developed. The further complication in simultaneously developing a housing development and ecotourism park for the London site meant that the timing of demolition and development had to be precise and closely adhered to. Development of the Clark County Wetlands Park could only have been accomplished in concert with a carefully formulated developmental schedule. Creation of the Nature Preserve and other site features, for instance, had to be arranged in coordination with construction of the erosion control measures and wash stabilization.

Water

For all regenerative landscape design projects involving water management (and almost all do), precise hydrological control is essential at all development stages. The success of the programs of aquatic revegetation at both the London Wetland Centre and the Clark County Wetlands Park, for example, depended on careful attention paid to regimes of flooding at all times. As well, because of its corporeal nature, water easily integrates both abuses and improvements across landscapes. It is necessary, therefore, to closely manage the element on the scale from individual drainage ditches and parking lots to large wetlands and lakes. The entire development of the London Wetland Centre, for example, was predicated upon the need to keep some water on-site at all times during development in order for the location to maintain its ecological status rating.

Contamination

Brownfields are a museum of past industrial development and contamination. As a result, site designs have to consider the issue of human exposure. This need not be an impossible constraint, however, that will stifle imaginative design. The Nature Preserve area at the Clark County Wetlands Park, for example, is still able to convey a site identity based on water without the need to bring visitors into bodily contact with the treated wastewater flowing in the Wash.

Engineering

Although there may be a wish to attempt to recover a degraded site solely through use of green or "soft" engineering approaches, realities of the situation may dictate otherwise. Often the past (and sometimes ongoing) abuses inflicted upon a landscape to bring it to its present state of degradation are severe enough that only through the employment of technologically advanced approaches and "hard" engineering structures will recovery occur. For the Clark County Wetlands Park, bioengineering techniques of vegetative erosion control would have been wholly inadequate on their own for withstanding the flows present in the Las Vegas Wash unless accompanied by riprap, gabions, steel pilings or concrete.

Recycling

The on-site recycling of material is extremely important to promoting a message about sustainable reuse. As such, what at first might be considered a developmental constraint can become a boon, providing both structural and nonstructural benefits. The "constraints" in having to reuse all the concrete from the reservoirs in construction of the London Wetland Centre became one of the most important educational benefits of practicing what is preached in terms of sustainable development. And certainly one of the more interesting stories to emerge from the regeneration work in the Clark County Wetlands Park was the reuse of the debris from imploded hotels on the Las Vegas Strip for erosion control.

Experimentation

Regenerative landscape design for recreation and ecotourism is a recently emerging field and as such should be approached by early practitioners with humility, not hubris, as a series of bold experimentations. The London Wetland Centre's trial-and-error investigations of fencing materials, planting techniques, and soil rebuilding, and of the Clark County Wetlands Park's attempts at various erosion-control measures provide valuable lessons about the honesty in admitting mistakes made in both cases.

Planting

It may be neither necessary nor desirable to completely replant the entire site intended for regeneration. Instead, the focused management of key "seed" areas will allow the naturally rejuvenating abilities of nature to do the rest of the work. Such a strategy becomes increasingly important with progressively larger areas such as the London Wetland Centre and particularly the Clark County Wetlands Park, where direct planting would have been cost-prohibitive. Another benefit in implementing this strategy is its educational role in informing about the resilience of nature in terms of self-healing and successional repair.

Regional context

Few things can be potentially more limiting and debilitating to the successful development and long-term sustainability of regenerative landscape design projects than the failure to situate the sites and recovery work in a regional context, both physically and conceptually. Good site-specific design alone is often insufficient to make a truly "great" regeneration project; careful attention must also be directed to regional planning. Efforts paid toward transportation concerns were instrumental to the London Wetland Centre becoming accepted by the local community as well as to attracting paying visitors to the site. Those planning the Clark County Wetlands Park soon realized that all the environmental problems of the Las Vegas Wash could never be solved solely by on-site interventions alone; rather, recommendations were made concerning the implementation of best management practices throughout the entire watershed, which would considerably improve the likelihood of success of the recovery actions undertaken within the park boundaries.

Adaptive management

Given the relative immaturity of the discipline of regenerative landscape design, problems will ensue and mistakes will be made. As a result, an upfront strategy of adaptive management is essential to be able to "expect the unexpected." Consequently, a wise approach is to proceed cautiously and have many alternatives on hand (or in mind) should they be needed. A conscious effort to play the "what if" game of assuming the likelihood of development "hiccups" and responses should be de rigueur. The noxious algal blooms that developed at the London Wetland Centre necessitated immediate alterations in water management in addition to the use of controlling agents that went against the modus operandi of the WWT but were deemed nonetheless essential to solve the problem. The extremely dynamic Las Vegas Wash necessitated repeated redesign of several of the erosion-control structures as well as the relocation of planned trails and other site features. In light of the heightened recognition about the process of change being the only real constant in the Las Vegas Wash, the multi-agency Comprehensive Adaptive Management Plan provides a blueprint for planning such uncertain futures.

Historical preservation and ecological rehabilitation

Naturalism

The imagined dichotomous schism between "nature" and "culture" is firmly put to rest in regenerative landscape design projects. Nature is no longer regarded as being something restricted to that which is outside of cities far removed from human influence. Instead the concept of "urban wilds" assumes importance. This occurs not only through the direct role of these recovered sites in sustaining (sub)urban wildlife, but also indirectly through informing and then inspiring urban dwellers about nature in such a way that they will become motivated to go to such areas and work toward preserving nature. Who would have thought that one of the world's top ecotourism destinations—the London Wetland Centre—could be found right in the middle of what was once the world's largest metropolis? And who would have ever imagined that Las Vegas—a city synonymous with sin and sprawl—could be the site of one of the most important wetland restoration projects in the world? This is certainly one of the most compelling lessons of success arising from these two case studies.

"Restoration"

Returning a landscape to some preconceived and imagined historic ideal is often untenable. In other words, the resulting artificiality of the "restored" landscape is by no means a detriment to judging the success of a regenerative landscape design project. Certainly the London Wetland Centre, with its dense mosaic of small and varied wetland habitats, in no way resembles the larger riparian floodplain marshes of various typologie that would have once existed on the site. And of course, the entire establishment of the Clark County Wetlands Park, built as it is around a system of rehabilitated wetlands in the Wash channel, is filled with irony in that the wetlands being "restored" were in fact created by humans in the first place through urban runoff. Regeneration projects such as these two case studies have really much more to do about "reinvention" than they do about "restoration" per se.

History and buildings

The most interesting (and some might argue, the most important) regenerative landscape design projects are those that pay homage to the previous industrial activities on the site rather than those that attempt to wipe the slate clean and disguise the fact that the location has had a rich cultural history. Much attention is spent, for example, on the Las Vegas Wash website in establishing a detailed timeline of human activities that have taken place there, elements of which are incorporated into the resulting design and program for the Clark County Wetlands Park.

People/hardscape–Nature/greenscape balance

Regenerative landscape design sites are not "natural." The more often that this simple truism is communicated and understood right from the very start, the fewer the controversies that will ensue about how the final recovered sites are interpreted and eventually judged by communities. Whereas a strong case can be made for a gentleness of landscape modification for greenfield projects, often the degree of abuse has been so severe for postindustrial brownfields or derelict agricultural landscapes that serious landscape reconfiguration may be the only solution to regenerate the site as a public park. In such situations, it is always important to remember that there is almost nothing a visiting public can inflict upon a reclaimed or regenerated landscape that is as severe an insult as the industrial or development process that brought the site to the abused brownfield status in the first place. There are few places where landscape designers can more appropriately dream large and act bold than when dealing with tortured manufactured sites. The balance achieved between the heavily artificial zoo-like portions of the London Wetland Centre and the more "natural" birding wetlands has been successful. The actions taken by the planners of the Clark County Wetlands Park to create a perimeter road for vehicles in addition to an hierarchical network of passive and active pedestrian trails offers another lesson in accepting the reality of the situation.

Environmental regulation

Sometimes strict adherence to environmental regulations, processes, and procedures can interfere with successful ecological rehabilitation, as the designers of the erosion-control structures in the Clark County Wetlands Park found. Consequently, establishing a good liaison with progressive thinkers within local regulatory and enforcement agencies can often be an important step toward bringing about site recovery.

Community participation

Bottom-up support

In some situations, the first ideas for regenerating and recovering a degraded site as a public park comes from within the local community rather than from a distant governmental agency. Such grass roots initiatives, as for example with respect to creation of the Clark County Wetlands Park, are instrumental for setting the process in motion, as well as for helping to ensure that the momentum doesn't wane over time. By mobilizing such support through developing an inclusive master plan with public participation, as this case study demonstrated, the government in turn gives ownership of the document back to the community, and as such increases the likelihood that the overall effort will succeed to the benefit of all.

Volunteerism

If the regenerated site is "good" in terms of being exciting, attractive, inspiring, and fun, and therefore capable of capturing the community's imagination and passion, then volunteers will be knocking on the door to help. Visitor centers, as both the London Wetland Centre and Clark County Wetlands Park found, can almost always rely upon being staffed by interested members from the local community. The key to sustaining volunteer support in such situations is to factor in a big element of education, for individuals are often looking for opportunities to be able to share their nature hobbies with the less-informed general public.

Engagement and stewardship

In harsh climates of financial drought, it becomes necessary to develop innovative ways in which to empower local communities by fostering participation in both the creation and especially the ongoing maintenance of regenerative landscape design projects. Once individuals become directly involved in the physicality of constructing or sustaining a regeneration site, they become more protective of it and are better able to defend it should support funding begin to wane. Regenerative landscape design can therefore be looked upon as a form of landscape gardening or repair, two activities that homeowners enthusiastically engage in on the small-scale. By having the public participate in the initial planting, both the London Wetland Centre and the Clark County Wetlands Park are assured of a long-time commitment of individuals to help with vegetation management. Further, creating opportunities for the public to actually build some of the trailside features or to "police" some sections of the site, as accomplished at the Clark County Wetlands Park, yet further ensures vigilant stewardship.

Vision keeper

The transformation of degraded landscapes into recreation parks or ecotourism centers may take considerable time, sometimes beyond the lifespan of any single organization or person. In such cases, some sort of umbrella "vision keeper" is required. The public-private-government triumvirate committee established to supervise work on the Las Vegas Wash is a good approach to sustain the momentum of regenerative landscape design particularly given the complicated political situation for that case study.

Education and ecotourism

Entertainment

Environmental education and entertainment need not be mutually exclusive. The London Wetland Centre is a striking demonstration that it is possible to blend functions of a zoo with those of a nature research park, without compromising the latter.

Visitor centers

Visitor centers are important for establishing site identity, as well as for educating about the processes of industrial history and site regeneration. To make the operation of such facilities financially self-sustaining, it may be necessary for them to be leased out for special events. The very high-end visitor center at the London Wetland Centre is clearly the central focus of the site, functioning both pragmatically, as for example, for offices, but also in terms of housing many of the interpretive operations and displays. Also, the gift shop here provides not only revenue but also a platform for education outreach.

Public knowledge

Because visitors to nature parks and regenerated landscapes may already possess a fair degree of environmental knowledge, the interpretive education program need not overly simplify things and talk down to the audience in nonchallenging ways. The London Wetland Centre soon found that its clientele (and it is important to note, that "clientele" is the correct word, since in this case a good deal of revenue is collected at the door) were already reasonably eco-literate. As such, environmental education programs should follow the recent trend of many popular animation movies in having elements appealing to both children and adults.

Sustainability

Regenerative landscape design is really a form of gardening writ large. With this in mind, the London Wetland Centre used the concept of sustainable gardens to teach visitors about wise land husbandry so that the overall environmental message from the recovered site could be adapted on a limited scale upon return to their homes.

Imagination

The most successful amenities developed at regenerative landscape design projects to attract and inform ecotourists are those that are truly imaginative. The way that the "airport for birds" observation gallery in the Visitor Centre at the London Wetland Centre is presented is a hallmark of success in how to connect people to nature. Likewise, the World Wetlands zoo-like display of clipped-winged exotic birds is a major drawing card to get people in the door where they can begin to be educated about the wider issues of wetland science and preservation.

Research

Regeneration landscape science, design, and management are relatively new activities that are engaged in without as large a background of information as their participants would hope for. To counter this paucity of data and established and proven methods, opportunities may exist to use the recovered landscapes as laboratories for actual environmental and ecological research. The numerous graduate student theses, informal investigations from neighborhood naturalist groups, and the ongoing in-house research program at the London Wetland Centre reflect a model about how this can be brought about.

Fun

Environmental education needn't always be steeped in doom, gloom, misery, and despair; rather it can be made fun. After all, perhaps the most compelling lesson arising from regenerative landscape design is a message of hope and renewal as something that is certainly worth celebrating. The London Wetland Centre well knows this. Their use of tongue-in-cheek displays and irrelevant, and humorous interpretive signage is one of the lasting memories that many take away from a visit there. As the Clark County Wetlands Park found, scheduling of special participatory events such as a variety of Earth Day activities and the like can be occasions for both education and play.

Impact

Perhaps one of the surest marks that distinguishes a truly great regenerative landscape design project from one that is merely good is the former's ability to have its message transcend the physical site by offering insight and instruction about how to repair or protect other landscapes. The distant improvements that have occurred as a result of influential, international visitors to the London Wetland Centre are extremely encouraging. The transference of solid ecological understanding about habitat management from the Clark County Wetlands Park to the entire American Southwest will be beneficial to the long-term survival of several endangered species there. Also, the many management lessons from the Clark County Wetlands Park have also been useful for providing advice on the restorative redevelopment of the devastated marshlands of southern Iraq.

Index